SpringerBriefs present concise summaries of cutting-edge research and practical applications across a wide spectrum of fields. Featuring compact volumes of 50–125 pages, the series covers a range of content from professional to academic.

Typical publications can be:

- A timely report of state-of-the art methods
- An introduction to or a manual for the application of mathematical or computer techniques
- A bridge between new research results, as published in journal articles
- A snapshot of a hot or emerging topic
- An in-depth case study
- A presentation of core concepts that students must understand in order to make independent contributions

SpringerBriefs are characterized by fast, global electronic dissemination, standard publishing contracts, standardized manuscript preparation and formatting guidelines, and expedited production schedules.

On the one hand, **SpringerBriefs in Applied Sciences and Technology** are devoted to the publication of fundamentals and applications within the different classical engineering disciplines as well as in interdisciplinary fields that recently emerged between these areas. On the other hand, as the boundary separating fundamental research and applied technology is more and more dissolving, this series is particularly open to trans-disciplinary topics between fundamental science and engineering.

Indexed by EI-Compendex, SCOPUS and Springerlink.

More information about this series at http://www.springer.com/series/8884

João M. P. Q. Delgado · António C. Azevedo ·
Ana S. Guimarães

Interface Influence on Moisture Transport in Building Components

The Wetting Process

 Springer

João M. P. Q. Delgado
Department of Civil Engineering
Faculty of Engineering
University of Porto
Porto, Portugal

António C. Azevedo
CONSTRUCT-LFC
University of Porto
Porto, Portugal

Ana S. Guimarães
CONSTRUCT-LFC
University of Porto
Porto, Portugal

ISSN 2191-530X   ISSN 2191-5318   (electronic)
SpringerBriefs in Applied Sciences and Technology
ISBN 978-3-030-30802-5   ISBN 978-3-030-30803-2   (eBook)
https://doi.org/10.1007/978-3-030-30803-2

This Springer imprint is published by the registered company Springer Nature Switzerland AG
The registered company address is: Gewerbestrasse 11, 6330 Cham, Switzerland

Preface

The knowledge of moisture migration inside building materials and construction building components is decisive for the way they behave when in use. The durability, waterproofing, degrading aspect and thermal behaviour of these materials are strongly influenced by the existence of moisture within their interior, which provoke changes in their normal performance, something that is normally hard to predict. Due to the awareness of this problem, the scientific community have performed various studies about the existence of moisture inside porous materials. The complex aspects of moisture migration phenomenon tended to encompass monolithic building elements, since the existence of joints or layers contributes to the change of moisture transfer along the respective building element that contributes to the change of mass transfer law. The presentation of an experimental analysis concerning moisture transfer in the interface of material that makes up masonry is described in such a way as to evaluate the durability and/or avoid building damages.

In this work, it was analysed, during the wetting process, the influence of different types of interface, commonly observed in masonry, such as perfect contact, joints of cement mortar, lime mortar, and the air space interface. The results allow the calculation of the hygric resistance. With these results, it is possible to use any advanced hygrothermal simulation program to study the water transport in building elements, considering different interfaces and their hygric resistance.

Porto, Portugal

João M. P. Q. Delgado
António C. Azevedo
Ana S. Guimarães

Contents

Chapter 1
Introduction

1.1 Motivation

Moisture damage is one of the most important factors limiting building performance. High moisture levels can damage construction (condensations, mould development) and inhabitants' health (allergic risks). As an example, rising damp coming from the ground that climbs through the porous materials constitutes one of the main causes of buildings degradation; especially the old ones (see Fig. 1.1).

Over the last decade, a great deal of concern has been placed on moisture in building envelopes due to the various problems caused. These problems include increased energy use, damage building envelopes, poor performance of heating, ventilation and air conditioning (HVAC) system, growth of mould, fungus and bacteria, and increased expense for building maintenance. In fact, moisture has been the largest factor, visible or invisible, limiting the service life of a building.

Furthermore, the knowledge of moisture transport is also essential for improving indoor air quality. Moisture accumulation in the building envelope can promote the growth of mold, mildew, and bacteria, which will grow anywhere [1]. A number of studies have found that mold and mildew in buildings can cause respiratory infections and increase the risk of asthma [2, 3]. In fact, mold is becoming one of the most important issues for building practices and, for example, Turner [4] presented a study in which a considerable amount of dollars spent on mold repairs and litigation in the United States of America was cited.

Another cause of moisture studies is the increased use of the building energy due to the presence of moisture in the building envelope, especially when evaporation and condensation occur inside them. In some cases, moisture accumulation may increase energy use by up to 25% [5]. Therefore, moisture studies are also important to the optimization in the use of the building energy.

Moisture transport through a building envelope normally involves interface phenomena, i.e., moisture transport across interfaces between building materials. Therefore, the knowledge of the interface phenomena is essential for the prediction

© The Author(s), under exclusive license to Springer Nature Switzerland AG 2020
J. M. P. Q. Delgado et al., *Interface Influence on Moisture Transport in Building Components*, SpringerBriefs in Applied Sciences and Technology,
https://doi.org/10.1007/978-3-030-30803-2_1

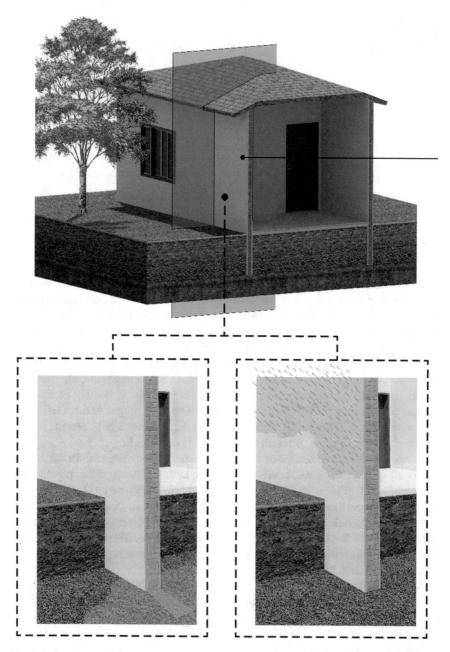

Fig. 1.1 Building wall in contact with water

of moisture behaviour in a building envelope. Most hygrothermal models treat materials as individual layers in perfect contact, i.e., the interface has no effect on the moisture transport. However, in practice, this might not always be true. Therefore, to appropriately evaluate the performance of a building envelope on moisture transport that lead to building envelope design guidelines, it is imperative to obtain a good understanding of the interface phenomena.

The major achievements of this work are the measurements of new experimental values of water absorption in samples of different building materials with and without joints at different positions (only a few experiments were presented in literature), new values of maximum transport flow, Q_{max}, function of the interface hygric resistance and a measurement and analysis of the moisture diffusion coefficient, D_w, of two building materials, using normal and anomalous diffusion models. In this study, this kind of model is analysed, since this study has particular relevance in hygrothermal numerical simulation and for evaluating the durability of building components. In literature it is possible to find a significant number of researchers [6–16] that have observed deviation from this behaviour when the infiltrating fluid is water and there is some potential for chemo-mechanical interaction with the material. For example, Küntz and Lavallée [6] discussed the anomalous behaviour and proposed a non-Fickian model as a more appropriate physical description.

1.2 Objectives and Methodology

The objectives of this work are the study of different interfaces: hydraulic contact, perfect contact and air space on moisture transport. To aim this objective, it is necessary to:

- Investigate the effects of different interfaces between building materials on moisture transport;
- Quantify hygric resistances in the interface between layers, which can be experimentally determined and is extremely important in relation to advanced hygrothermal simulation programmes;
- Investigate the effects of bonding on moisture transport.

References

1. B. Flannigan et al., Health implications of fungi in indoor environments—an overview, in *Health Implications of Fungi in Indoor Environments, Air Quality Monographs*, ed. by R. Samson, vol. 2 (1994) , pp. 3–28
2. O.M. Koskinen et al., The relationship between moisture or mould observations in houses and the state of health of their occupants. Eur. Respir. J. **14**(6), 1363–1367 (1999)
3. S. Armstrong, The fundamentals of fungi. ASHRAE J. **44**(11), 18 (2002)
4. F. Turner, Moisture and mold. ASHRAE J. **44**(11), 4 (2002)

5. ASHRAE Handbook et al., *American Society of Heating*. Refrigeration and Air-Conditioning Engineers, Inc. (2001)
6. M. Küntz, P. Lavallée, Experimental evidence and theoretical analysis of anomalous diffusion during water infiltration in porous building materials. J. Phys. D Appl. Phys. **34**(16), 2547 (2001)
7. W. Woodside, Calculation of the thermal conductivity of porous media. Can. J. Phys. **36**(7), 815–823 (1958)
8. N. Mendes, P.C. Philippi, A method for predicting heat and moisture transfer through multilayer walls based on temperature and moisture content gradients. Int. J. Heat Mass Transf. **48**, 37–51 (2005)
9. M. Bomberg, *Moisture Flow Through Porous Building Materials*. Division of Building Technology, Lund Institute of Technology, Lund, Sweden, Report no. 52 (1974)
10. H. Janssen, H. Derluyn, J. Carmeliet, Moisture transfer through mortar joints: a sharp-front analysis. Cem. Concr. Res. **42**, 1105–1112 (2012)
11. P. Mukhopadhyaya, P. Goudreau, K. Kumaran, N. Normandin, Effect of surface temperature on water absorption coefficient of building materials. J. Therm. Envelope Build. Sci. **26**, 179–195 (2002)
12. S. Roels et al., Interlaboratory comparison of hygric properties of porous building materials. J. Therm. Environ. Build. **27**(4), 307–325 (2004)
13. N.A. García et al., Propiedades físicas y mecánicas de ladrillos macizos cerámicos para mampostería. Cien. Ing. Neogranadina **22**(1), 43–58 (2012)
14. V.P. Freitas, Moisture transfer in building walls—interface phenomenon analyse. Ph.D. thesis, Faculty of Engineering, University of Porto, Porto, Portugal (1992)
15. P. Crausse, Fundamental study of heat and moisture transfer in unsaturated porous medium. Ph.D. thesis, ENSEEIHT, Toulouse, France (1983)
16. B. Perrin, Study of coupled heat and mass transfer in consolidated, unsaturated porous materials used in civil engineering. Ph.D. thesis, University Paul Sabatier (1985)

Chapter 2
State-of-the-Art

2.1 Introduction

Transport phenomena in porous media occurs in diverse fields of science and engineering, ranging from agricultural, biomedical, building, ceramic, chemical, and petroleum engineering to food and soil sciences. Morrow [1] provides an extensive description of the problems involving porous media. For building engineering, obtaining a good understanding of moisture transport in building envelopes is becoming one of the most important tasks. In the last few decades, many studies investigating moisture transport in building envelopes have been published, which have helped to improve overall building envelope design. This chapter presents a brief review of these studies.

2.2 Moisture and Interface Is Building Components

2.2.1 Moisture Transport in Buildings

Many building materials are porous media [2], consisting of an extremely complicated network of channels and obstructions [3]. For all porous media, part of the domain is a solid matrix, which is occupied by a persistent solid phase; the remaining part is void space, which is either occupied by a single fluid phase, or by a number of phases [4].

Moisture may exist in building materials in all three states, i.e., solid (ice), liquid (water), and gas (water vapour). However, it is difficult to experimentally distinguish the different physical states of moisture. Furthermore, the ratio of the different states of moisture varies in natural conditions.

© The Author(s), under exclusive license to Springer Nature Switzerland AG 2020
J. M. P. Q. Delgado et al., *Interface Influence on Moisture Transport in Building Components*, SpringerBriefs in Applied Sciences and Technology, https://doi.org/10.1007/978-3-030-30803-2_2

To understand the moisture behaviour of a specific building material, it is necessary to know the moisture storage characteristics of the material. There are two typical approaches to describing these characteristics. One states that moisture storage consists of three regimes: diffusion regime, transition regime, and capillary regime [4]. The diffusion regime refers to a state below the maximum sorption state occurring at 95%RH. The transition regime refers/points to a transition state between diffusion regime and capillary regime. The capillary regime starts when the liquid transport is continuous. However, another more sound approach divides moisture storage in a building material into three regions: sorption moisture or hygroscopic region, capillary water region, and supersaturated region [5, 6]. The sorption region refers to moisture storage in a range between 0%RH and 95%RH and includes all moisture content resulting from water vapour sorption. From 95%RH to free water saturation, it is capillary water region. In this region, moisture storage is characterized by capillary suction. Supersaturated region ranges from free water saturation to the maximum saturation. The concept hygroscopic moisture indicates that a certain experimental method is used [7]. As a consequence, a practical limit for sorption region may range from 95%RH to 98%RH, which depends on experimental facilities [7]

Moisture storage in a material depends on the history of drying and wetting of the material, i.e., hysteresis effect [8]. The hysteresis effect may be due to: (a) the rain-drop effect; (b) the ink-bottle effect; (c) the air entrapment effect.

The rain-drop effect refers to the difference between advancing and receding contact angles. The rain-drop effect is probably caused by the contamination of fluid, solid or surface roughness [9]. The ink-bottle effect refers to the different stable configuration of water resulting from drainage and imbibition. The ink-bottle effect can be attributed to the fact that the wide range of pores with different dimensions in materials may result in different stable configurations [3]. The air entrapped in a material after draining or rewetting may also cause hysteresis [10].

However, for most building materials, the hysteresis effect is not significant and, therefore, sorption curves are normally used to characterize building materials in the hygroscopic region [11]. For those building materials with a significant hysteresis effect, Pedersen [9] concluded that the average of sorption and desorption curves can be used to characterize the moisture storage characteristics.

Moisture can be transported in building materials in the form of water vapour, liquid water and adsorbate film. For the water vapour transport, the thermal diffusion resulting from the gradient of temperature, i.e., the Soret effect, approximately accounts for 0.05% of the overall moisture transport at normal conditions to which building structures are exposed [6]. In addition, the air pressure around constructions is difficult to determine [9]. Therefore, many studies of water vapour transport mainly investigate the water vapour diffusion resulting from the gradient of partial water vapour pressure, including pure diffusion, Knudsen diffusion or effusion, and mixed diffusion [12]. Pure diffusion is dominated by the interactions between gas molecules, while the Knudsen diffusion or effusion is dominated by the interactions between gas molecules and pore walls. The mixed diffusion consists of both types of diffusion.

Surface diffusion refers to a phenomenon in which the water vapour diffusion increases as relative humidity increases. Surface diffusion may result from the fact that when building materials are in contact with moist air, water molecules are localized on the inner surfaces and form a water film. Since water molecules are more mobile in a thicker film than a thinner one, water molecules will move from relatively thick places to relatively thin places [13–15] suggested that the Van der Waals forces and the interactions between water dipoles and solid surfaces may attract water in films. References [16, 17] concluded that the surface diffusion could be implicitly calculated by an overall moisture diffusivity coefficient and the gradient of moisture content.

In summary, water can be transported in various forms, e.g., hydraulic flow, capillary flow, and gravitational flow in building materials. However, some of them, such as hydraulic flow and electrokinesis, are negligible for most buildings [6]. Therefore, capillary flow is a major concern of many moisture studies. The driving force of capillary flow is capillary suction, which is an equilibrium property and is directly related to surface tension, i.e., interfacial potential energy [1]. Capillary suction is defined as the difference between the pressure for the non-wetting phase and the pressure for the wetting phase [18], expressed in Eq. (2.1):

$$P_c = P_{nW} - P_W \tag{2.1}$$

In Eq. (2.1), P_c is the capillarity suction, P_{nW} is the pressure of the non-wetting phase and P_W is the pressure of the wetting phase (all in Pa).

2.2.2 Numerical Methods

To study the water transport in building materials and components is essential to use numerical simulations that can easily predict the real behaviour; these models can be split into two types:

- Continuum models;
- Discrete models.

Continuum modelling is a classical engineering approach to describe a system using macroscopic equations and suitable effective transport properties [19]. In contrast to continuum models, the strategy of discrete models is to disassemble pore space into discrete elements and to reassemble these elements into a model that preserves the basic geometrical and topological features of the pore space [20–22]. Therefore, discrete models are helpful in understanding the physics of moisture flow in porous media. However, the main drawback of discrete models is that, from a practical point of view, excessively large computational efforts are required for a realistic discrete treatment of a system [18]. Moreover, transport problems cannot be formulated and solved at a microscopic level due to the lack of information concerning the microscopic configuration of the interphase boundaries. Such a solution is usually of no interest in practice [4].

Compared to discrete modelling, continuum modelling has been widely applied because of its convenience and sufficient accuracy. The main advantages of continuum models are:

- Avoid specifying the exact configuration of the inter-phase boundaries;
- Describe a transport process in terms of differentiable quantities, which can be solved by mathematical analysis;
- Measure the macroscopic quantities used, which is useful in solving the field problems of practical interest [4].

Most moisture models are based on Fick's law, Darcy's law and Richards' equation. One of the main differences between these models is the driving potential utilized. In contrast to heat transfer, in which the temperature is recognized as the potential, unanimity on driving forces is lacking in moisture models. For example, Philip et al. [23], Luikov [24] and De Vries [25] used moisture content as a potential, Kießl [26] used a redressed potential generated by relative humidity and radius of pores [8]. Salonvaara [27] used moisture content and water vapour pressure as the potentials. Matsumoto [28] defended temperature and chemical potentials. Kunzel et al. [11] argued that relative humidity is the potential. Burch et al. [29] used capillary suction and water vapour pressure.

In 2003, a review report of hygrothermal models for building envelope retrofit analysis made by Canada Mortgage and Housing Corporation has identified 45 hygrothermal modelling tools, and in the last years, new hygrothermal models were developed, most of them during Annex 41 [30]. However, most of the hygrothermal models available in the literature are not readily available to the public outside the organization where they were developed. In fact, only a reduced number of hygrothermal modelling tools are available to the public in general.

DELPHIN 5 and the WUFI, two of the models cited in this study, are the more well-known and are available at the FEUP (Laboratory of Building Physics of Faculty of Engineering of University of Porto) where this work is being developed, and are utilized for the transport of liquid water, which is part of the objective of this study. Additionally, other important models are cited because the value of the references they contribute in relation to material properties and boundary conditionals outside.

DELPHIN 5 (commercial program)—a one or two-dimensional model for transport of heat, air, moisture, pollutant and salt transport in porous building materials, assemblies of such materials and building envelopes in general. The Delphin program can be used to simulate transient mass and energy transport processes for arbitrary standard and natural climatic boundary conditions (temperature, relative humidity, driving rain, wind speed, wind direction, short and long wave radiation). This simulation tool is used for: (1) Calculation of thermal bridges, including evaluation of hygrothermal problem areas (surface condensation, interstitial condensation); (2) Design and evaluation of inside insulation systems; (3) Evaluation of ventilated façade systems, ventilated roofs; (4) Transient calculation of annual heating energy demand (under consideration of moisture dependent thermal conductivity);

(5) Drying problems (basements, construction moisture, flood, etc.); (6) Calculation of mould growth risks and further applications.

A particular advantage of the numerical simulation program is the possibility to investigate variants concerning different constructions, different materials and different climates. Constructive details of buildings and building materials can be optimised using the numerical simulation, and the reliability of constructions for different given indoor and outdoor climates can be evaluated. A large number of variables, such as moisture contents, air pressure, salt concentrations, temperatures, diffusive and advective fluxes of liquid water, water vapour, air, salt, heat and enthalpy which characterize the hygrothermal state of building constructions, can be obtained as functions of space and time [31].

The governing equations for moisture mass balance, air mass balance, salt mass balance and internal energy balance are, respectively:

$$\frac{\partial}{\partial t}\left(\rho_w\theta_l + \rho_v\theta_g\right) = -\frac{\partial}{\partial x}\left[\left(\rho_w/v - j_{disp} - j_{diff}\right)\theta_l + \left(\rho_v/v + j_{diff}\right)\theta_g\right] \quad (2.2)$$

$$\frac{\partial}{\partial t}\left(\rho_a\theta_g\right) = -\frac{\partial}{\partial x}\left[\left(\rho_a/v - j_{diff}\right)\theta_g\right] \quad (2.3)$$

$$\frac{\partial}{\partial t}\left(\rho_s\theta_l + \rho_p\theta_p\right) = -\frac{\partial}{\partial x}\left[\left(\rho_s/v + j_{disp} + j_{diff}\right)\theta_l\right] \quad (2.4)$$

$$\frac{\partial}{\partial t}\left[\rho_m c_{pm}T + \rho_p c_{pp}T\theta_p + \rho_l c_{pl}T\theta_l + \left(\rho_v c_{pv}T + \rho_a c_{pa}T\right)\theta_g\right]$$
$$= -\frac{\partial}{\partial x}\left[\rho_l c_{pl}T/v\theta_l + \left(\rho_v c_{pv}T + \rho_a c_{pa}T\right)/v\theta_g\right] -$$
$$- \frac{\partial}{\partial x}\left[-\lambda\frac{\partial T}{\partial x} + (h_s - h_w)\left(j_{disp} + j_{diff}\right)\theta_l + (h_v - h_a)j_{diff}\theta_g\right] \quad (2.5)$$

where ρ_v (kg m^{-3}) is the partial moisture density, φ (−) the relative humidity, δp (s) the water vapor permeability, p_s (Pa) the partial pressure of saturated water vapor in the air, H (J m^{-3}) the enthalpy density, L_v (J kg^{-1}) the latent heat of evaporation of water, λ (W m^{-1} K^{-1}) the thermal conductivity, T (K) the temperature and D_φ (m^2 s^{-1}) the liquid water transport coefficient.

WUFI (commercial program)—a one or two-dimensional model for heat and moisture transport developed by Fraunhofer Institute in Building Physics (IBP) and validated using data derived from outdoor and laboratory tests. This model allows for realistic calculation of the transient hygrothermal behaviour of multi-layer building components exposed to natural climate conditions [32]. Heat transfer occurs by conduction, enthalpy flow (including phase change), short-wave solar radiation and long-wave radiative cooling (at night). Convective heat and mass transfer is not modelled. Vapour-phase transport is by vapour diffusion and solution diffusion, and liquid-phase water transport is by capillary and surface diffusion. The governing equations for moisture and energy transfer are, respectively:

$$\frac{\partial w}{\partial \varphi} \frac{\partial \varphi}{\partial t} = \nabla \left(D_\varphi \nabla \varphi + \delta_p \nabla (\varphi p_{sat}) \right) \tag{2.6}$$

$$\frac{\partial H}{\partial T} \frac{\partial T}{\partial t} = \nabla (\lambda \nabla T) + h_v \nabla \left(\delta_p \nabla (\varphi p_{sat}) \right) \tag{2.7}$$

and the water vapour diffusion resistance factor, μ, used by WUFI is given by:

$$\mu = \frac{\delta_a}{\delta_p} = \frac{2.0 \times 10^{-7} T^{0.81} / P_n}{\delta_p}. \tag{2.8}$$

where ρ_v (kg m^{-3}) is the partial moisture density, φ ($-$) the relative humidity, δp (s) the water vapor permeability, p_s (Pa) the partial pressure of saturated water vapor in the air, H (J m^{-3}) the enthalpy density, L_v (J kg^{-1}) the latent heat of evaporation of water, λ (W m^{-1} K^{-1}) the thermal conductivity, T (K) the temperature and D_φ (m^2 s^{-1}) the liquid water transport coefficient.

The programs available for the public in general were analysed in detail [30, 33, 34], namely the input of material properties and the boundary conditions (inside and outside). The most important exterior environmental loads (influence directly the transport of moisture) are: (1) ambient temperature; (2) ambient relative humidity; (3) solar diffuse; (4) solar direct; (5) cloud index; (6) wind velocity; (7) wind orientation and (8) horizontal rain precipitation.

From these simulations programs, the most important ones for this work, whose purpose is to study the transport of moisture in the liquid phase, are those that require the property Liquid Diffusivity and Boundary Conditions (outside) Precipitation.

Finally, as the purpose of most hygrothermal models is usually to provide sufficient and appropriate information needed for decision-making, it is suggested four items that should be used when modelling a single component of the building envelope or a multizonal building: (1) the software must be available in the public domain (freeware or commercially); (2) the software must be suitable for the single component or a multizonal building analysis under consideration; (3) the programs must be of reasonably recent vintage or with recent updated development and (4) the software must be "user-friendly".

As the programs have different hygrothermal potentialities, strengths and weaknesses, such as the availability to modelling the heat and moisture transfer by air movement, 2-D or 3-D phenomena, or the capability to simulate high number of zones in a reasonable execution time, the investigators need to select the hygrothermal simulation tools that are better suited to their requirements.

2.2.3   Interface Influence on Moisture Transport

Rising damp consists of the migration by the capillary movement of water from the soil in the building walls through the porous network of the materials used in the walls and building floors. This type of moisture is more expressive in old buildings, mostly made of masonry, in which porous materials such as red clay bricks and mortars were used and whose protection against this phenomenon is almost non-existent.

Incident rain is one of the main factors of humidification of façades. To quantify incident rain is quite complex and depends on several parameters, such as building geometry, façade position, topography, wind speed, wind direction, horizontal precipitation intensity and drops distribution [35].

On a vertical surface, the action of the incident rain results from the combination of rain and wind is one of the main sources of façade moisture. Atmospheric parameters, wind and rain, change greatly in time and space, so the amount of incident rain on a façade also presents great variability [36]. Nevertheless, the determination of the amount of incident rain on a façade is fundamentally carried out using semi-empirical methods, namely, an analytical relationship that defines the intensity between rain and the movement of the velocity perpendicular to the phase by the intensity of the precipitation. Nore et al. [36] indicates that the proportionality factor, called the incident rainfall coefficient, Rci, is a function of characteristics and of the building itself, and can be determined from experimental results or simulations with computational models of fluid mechanics (CFD).

According to Abuku [37], incident rain absorption on the façade depends on two consecutive boundary conditions. The absorbed moisture flow equals the amount of incident rainfall if the façade is not completely wet, i.e. when the moisture content at the surface is less than the moisture content parallel to the capillary saturation. When capillary saturation is reached, i.e. excess water flows down the vertical surface, the incident rain is no longer considered and ensures relative moisture of 100%. The absorbed moisture flow depends only on capillary pressure and surface saturation pressure.

Masonry is the prevalent solution for constructing building envelopes. While its constituent materials (bricks and cement mortar) are reasonably well understood, the hygric behaviour of masonry is often shown to deviate from the normal unsaturated flow theory. In literature, some researchers refer to an imperfect contact and, hence an interface resistance in the brick-mortar bond plane.

A building wall, in general, consists of multiple layers (see Fig. 2.1), and thus the investigation of the moisture transfer provides knowledge about the continuity between layers.

All of these phenomena enable the retarded water uptake in brick–mortar composites found in the experimental investigations to be explained [38]. Due to its complexity, the incorporation in numerical models of the retarded water uptake across a brick–mortar interface is often done by the implementation of the phenomena mentioned [39]. In literature, different approaches were found. Derluyn et al. [38] obtained an agreement between numerical and experimental investigations by assuming a perfect hydraulic interface contact in combination with modified mortar properties. In contrast, Qiu [40] neglected the change in material properties and examined the liquid transport across the interface between aerated concrete and mortar, and simulated the moisture behaviour by the use of an interface resistance. According to

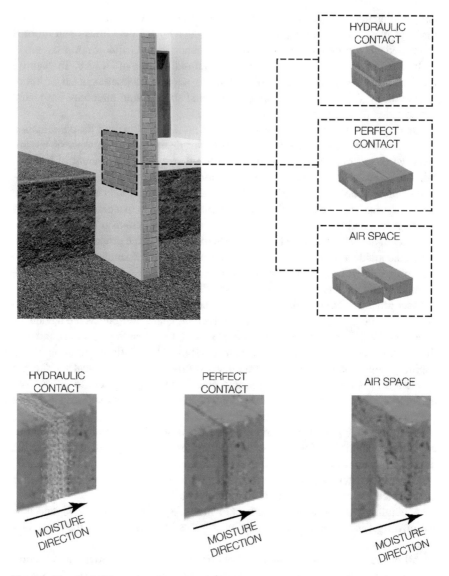

Fig. 2.1 Sketch of different interfaces observed in building components

the authors, this simplification is acceptable if the interface resistance is determined after capillary saturation of the first layer.

Derluyn et al. [38] took into account an interface resistance as well as a change in mortar properties. The authors determined the brick–mortar interface resistance based on the water uptake from the brick layer into the mortar joint. Hence, the values were determined for the liquid transport into a material with a lower absorption. Besides the mortar properties, the interface resistance was also found to depend on the curing conditions. For a dry cured mortar, a higher interface resistance was obtained compared to the wet cured composite. Similar interface resistances analytically obtained by Janssen et al. [41] prove the validity of Derluyn's approach.

Although several studies concerning the liquid transport in multilayered composites can be found [42–47], currently only a limited number of values for the interface resistance in multilayered composites are available [42, 43, 47]. These values were determined based on the moisture profiles measured during an imbibition's experiment and can be strongly case-dependent. For instance, the mortar type (e.g. W/C factor, additives), the brick type (e.g. capillary), the curing conditions (e.g. moisture content of the brick), the thickness of the mortar joint, etc. may have a potential impact on the interface resistance and the modification of the material properties.

Freitas et al. [47] described three kinds of continuity between layers: "Hydraulic contact" when there is an interpenetration of both layer's porous structure; "Perfect contact" when there is a contact without interpenetration and "Air space" between layers when there is an air box of a few millimeters wide between the layer's porous structure (see Fig. 2.1).

References

1. N.R. Morrow, S. Ma, X. Zhou, X. Zhang, Characterization of wettability from spontaneous imbibition measurements, in *Proceedings of the 45th Annual Technical Meeting of the Petroleum Society of the CIM*, Calgary, Alberta, Canada, 12–15 June 1994
2. M.K. Kumaran, *Moisture diffusivity of spruce specimen, IEA, Annex 24, HAMTIE* (International Energy Agency, New York, 1992)
3. R.A. Greenkorn, *Flow phenomena in porous media* (Marcel Dekker, New York and Basel, 1983)
4. J. Bear, Y. Bachmat, *Introduction to Modeling of Transport Phenomena in Porous Media* (Springer, Berlin, 1990), 305 pages
5. D.M. Burch, Controlling moisture in the roof cavities of manufactured housing. NISTIR 4916, USA (1992)
6. H.M. Künzel, Th. Großkinsky, Feuchtebelastungen beeinträchtigen die Wirkung von Dampfbremspappen. Fraunhofer-Institut für Bauphysik (1997)
7. K.H. Bomberg, Rezeptive Musiktherapie im Rahmen integrativer Psychotherapie. Ph.D. thesis, Verlag nicht ermittelbar, Germany (1989)
8. H. Hens, IEA Annex 24: Heat, air and moisture transfer through new and retrofitted insulated envelope parts (HAMTIE). Final Report, Molenda (1996)
9. A.W. Adamson, *Physical Chemistry of Surfaces*, 5th edn. (Wiley, New York, 1990)
10. A. Poulovassilis, Hysteresis of pore water, an application of the concept of independent domains. Soil Sci. **93**(6), 405–412 (1962)
11. H. Künzel, *Simultaneous Heat and Moisture Transport in Building Components—One and Two-Dimensional Calculation using Simple Parameters* (IBP Verlag, Stuttgart, Germany, 1995)
12. C. Rode, Combined heat and moisture transfer in building constructions. Thermal Insulation Laboratory, Technical University of Denmark (1990)
13. P.C. Carman, *Flow of Gases Through Porous Media* (Academic Press, New York, USA, 1956)
14. L. Pel, Moisture transport in porous building materials. Ph.D. thesis, Technical University Eindhoven, The Netherlands (1995)
15. B.M. Parker, Some effects of moisture on adhesive-bonded CFRP-CFRP joints. Compos. Struct. **6**(1–3), 123–139 (1986)

16. M.K. Kumaran, G.P. Mitalas, M. Bomberg, Fundamentals of transport and storage of moisture in building materials and components. Moisture Control Build. **18**, 3–17 (1994)
17. J. Straube, E. Burnett, Overview of hygrothermal (HAM) analysis methods. ASTM Manual **40**, 81–89 (2001)
18. F.A.L. Dullien, *Porous Media: Fluid Transport and Pore Structure* (Academic Press, New York, USA, 1992)
19. M. Sahimi, Fractal and superdiffusive transport and hydrodynamic dispersion in heterogeneous porous media. Transp. Porous Media **13**(1), 3–40 (1993)
20. M. Sahimi, Nonlinear transport processes in disordered media. AIChE J. **39**(3), 369–386 (1993)
21. F. Descamps. Continuum and discrete modelling of isothermal water and air transfer in porous media. Ph.D. Thesis, Catholic University of Leuven, Leuven, Belgium (1997)
22. I. Fatt, *The Network Model of Porous Media* (Society of Petroleum Engineers, USA, 1956)
23. J.R. Philip, D.A. de Vries, Moisture movement in porous materials under temperature gradients. Eos Trans. Am. Geophys. Union **38**(2), 222–232 (1957)
24. A.V. Luikov, Heat and mass transfer in capillary-porous bodies. Adv. Heat Transf. **1**, 123–184 (1964)
25. D.A. de Vries, J.R. Philip, Soil heat flux, thermal conductivity, and the null-alignment method 1. Soil Sci. Soc. Am. J. **50**(1), 12–18 (1986)
26. K. Kießl, Kapillarer und dampfförmiger Feuchtetransport in mehrschichtigen Bauteilen: Rechnerische Erfassung und bauphysikalische Anwendung. Ph.D. thesis, Universität-Gesamthochschule Essen, Germany (1983)
27. M. Salonvaara, *Moisture Potentials: Numerical Analysis of Two Differential Equations* (Internal Reports, USA, 1993)
28. K. Matsumoto et al., Solidification of porous medium saturated with aqueous solution in a rectangular cell. Int. J. Heat Mass Transf. **36**(11), 2869–2880 (1993)
29. D.A. Burch, J. Chi, MOIST: a PC program for predicting heat and moisture transfer in building envelopes: Release 3.0. US Department of Commerce, National Institute of Standards and Technology, USA (1997)
30. J.M.P.Q. Delgado, N. Ramos, E. Barreira, V.P. Freitas, A critical review of hygrothermal models used in porous building materials. J. Porous Media **13**(3), 221–234 (2010)
31. A. Nicolai, J. Zhang, J. Grunewald, Coupling strategies for combined simulation using multizone and building envelope models, in *Proceedings of Building Simulation (BS)*, Beijing, China, 3–6 Sept 2007
32. H.M. Künzel, K. Kießl, Moisture behaviour of protected membrane roofs with greenery, in *CIB W40 Meeting*, Prague, 30 Aug–3 Sept 1999
33. N.M.M. Ramos, J.M.P.Q. Delgado, E. Barreira, V.P. de Freitas, Hygrothermal numerical simulation: application in moisture damage prevention, in *Numerical Simulations—Examples and Applications in Computational Fluid Dynamics*, vol. 1, Chap. 6, ed. L. Angermann (INTECH Publishers, 2010), pp. 97–122
34. J.M.P.Q. Delgado, E. Barreira, N.M.M. Ramos, V.P. de Freitas, *Hygrothermal Numerical Simulation Tools Applied to Building Physics* (Springer, Germany, 2012)
35. A.S. Freitas, Avaliação do comportamento higrotérmico de revestimentos exteriores de fachadas devido à acção da chuva incidente. M.Sc. thesis, Faculdade de Engenharia da Universidade do Porto, Portugal (2011)
36. K. Nore, B. Blocken, B.P. Jelle, J.V. Thue, J. Carmeliet, A dataset of wind-driven rain measurements on a low-rise test building in Norway. Build. Environ. **42**, 2150–2165 (2007)
37. M. Abuku et al., Numerical Simulation of absorption and evaporation of Wind driven rain at buildings facades, in *12th Symposium for Building Physics*, vol. 1 (Technishe Universitat Dresden, Dresden, 2007), pp. 588–595
38. H. Derluyn, H. Janssen, J. Carmeliet, Influence of the nature of interfaces on the capillary transport in layered materials. Constr. Build. Mater. **25**(9), 3685–3693 (2011)
39. R.J. Gummerson, C. Hall, W.D. Hoff, R. Hawkes, G.N. Holland, W.S. Moore, Unsaturated water flow within porous materials observed by NMR imaging. Nature **281**, 56–57 (1979)

40. X. Qiu, Moisture transport across interfaces between building materials. Ph.D. thesis, Concordia University, Montreal, Canada (2003)
41. H. Janssen, H. Derluyn, J. Carmeliet, Moisture transfer through mortar joints: a sharp-front analysis. Cem. Concr. Res. **42**, 1105–1112 (2012)
42. J.M.P.Q. Delgado, A.S. Guimarães, V.P. de Freitas, I. Antepara, V. Kočí, R. Černý. Salt damage and rising damp treatment in building structures. Adv. Mater. Sci. Eng. 2016, Article number ID 1280894, 13 pages (2016)
43. A.S. Guimarães, J.M.P.Q. Delgado, V.P. Freitas, Rising damp in walls: evaluation of the level achieved by the damp front. J. Build. Phys. **37**, 6–27 (2013)
44. A.S. Guimarães, J.M.P.Q. Delgado, V.P. Freitas, Rising damp in building walls: the wall base ventilation system. Heat Mass Transf. **48**, 2079–2085 (2012)
45. V.P. Freitas, A.S. Guimarães, J.M.P.Q. Delgado, The HUMIVENT device for rising damp treatment. Recent Pat. Eng. **5**, 233–240 (2011)
46. V.P. de Freitas, Moisture transfer in building walls—interface phenomenon analysis. Ph.D. thesis, Faculdade de Engenharia da Universidade do Porto, Porto, Portugal (1992)
47. V.P. de Freitas, V. Abrantes, P. Crausse, Moisture migration in building walls: analysis of the interface phenomena. Build. Environ. **31**, 99–108 (1996)

Chapter 3
Moisture Content Determination

3.1 Introduction

Building materials are, for the most part, porous structures which generate permanent changes when in contact with moisture, which could be in the vapour phase and in the liquid phase, with the environment in which they are placed. This behaviour conditioned their properties and their durability. It should be noted that humidity is a major cause of the pathologies observed in buildings. The profile of the transient moisture content describes moisture content as a function of time and place. This profile is essential for determining the moisture diffusivity of a material and is also important for studying the interface phenomena. In this sub-chapter, a brief description of the most important experimental transient methods—Gravimetric Method, Nuclear Magnetic Resonance (NMR) method, X-Ray Analysis and Gamma Ray attenuation method—to determine moisture content profiles is going to be presented.

3.2 Gravimetric Method

The Gravimetric Method is a classic method to determine the transient moisture content [1]. This method consists of the selection of a sample material, weighed before and after the drying process. This method permits calculating, by mass difference, the amount of water contained inside the material.

Of the various techniques available, the gravimetric method is the simplest since ISO 15148 [2] only provides for vertical absorption tests. This water absorption test is measured using prismatic geometry specimens of the analysed material under prescribed atmospheric pressure conditions. After drying to constant mass, one side of the specimen is immersed in 5–10 mm of water for a specific period of time and the mass increase is determined (see Fig. 3.1).

© The Author(s), under exclusive license to Springer Nature Switzerland AG 2020
J. M. P. Q. Delgado et al., *Interface Influence on Moisture Transport
in Building Components*, SpringerBriefs in Applied Sciences and Technology,
https://doi.org/10.1007/978-3-030-30803-2_3

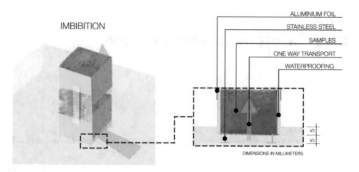

Fig. 3.1 Imbibition model used in the research

Although it is a direct and apparently accurate technique, the gravimetric method has the disadvantage of being destructive. It should be noted that the cutting of the consolidated materials, necessary for observation of the wet front in the specimen, produces heat and consequently a disruption in the distribution of moisture.

3.3 Nuclear Magnetic Resonance (NMR) Method

The NMR method is a non-destructive method and is based on the fact that, hydrogen nuclei in building materials occur only in the form of water. An extensive description of the NMR method can be found in Refs. [3–5]. The neutron scanning method is also a non-destructive method. When a beam of neutrons passes through a material and neutrons interact with the nuclei of the material. As a consequence, neutrons are scattered, slow down or diffuse, which results in thermal neutrons having altered the direction of travel and reduced energy. Due to the large scattering cross-section of hydrogen, the intensity of neutrons significantly depends on the amount of water. Moisture content of a material can thus be determined by comparing the intensity of neutrons through the material at dry state with the intensity at wet state [1, 6].

3.4 X-Ray Analysis

The water absorption by capillarity can be analyzed by the X-Ray projection method, as demonstrated Janssen et al. [7]. The sample is placed in contact with water and the moisture transfer perpendicular to the mortar joint is evaluated.

The X-Ray images are obtained at predefined time intervals in each assay by subtracting the X-ray image from the sample results, initially dried on an X-ray image after water absorption (see Fig. 3.2).

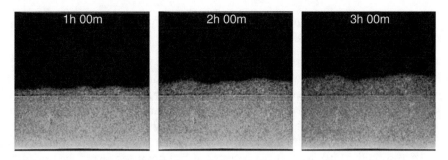

Fig. 3.2 Example of gamma-ray image

3.5 Gamma Ray Attenuation Method

The gamma ray attenuation method is one of the most widely used non-destructive methods to determine the transient moisture content profile. When gamma rays pass through a material, the adsorption and scattering of gamma rays depend on the nature of the material. Because the absorption of gamma rays varies with moisture in the material, moisture content of a material can be determined by comparing the intensity of gamma rays through the material at dry state with the material at wet state. A detailed description of the gamma ray attenuation method can be found in Refs. [8–11].

In order to study the imbibition and drying kinetics and to determine the water diffusivity coefficients of the various building materials, the LFC-FEUP has a gamma radiation attenuation moisture measuring device. These are non-destructive tests where it is possible to calculate the moisture content profiles in a continuous absorption or drying process.

The laboratory tests were done with gamma ray equipment that already had been built impressed and recently adapted to that propose using an automatic system. Experimental measurement of moisture content profiles in multilayer specimen's subjected to humidifying and drying of liquid water allows for the validation of a future mathematical model that will be developed, and giving important information about the influence of salts solutions as well as the different layers in the moisture transport process. That objective includes basic research, as well as measuring moisture content profiles with a large laboratory experimental campaign supported in gamma ray equipment. It was necessary to adapt the existing prototype to this specific task. Gamma ray equipment was adapted to profile with automatic measurements and linked to a computer, allowing for an almost continuous moisture content registration throughout the specimen.

The absorption and drying tests to study the moisture transfer in different materials were done with LFC-FEUP samples. This was done to simulate absorption and drying cycles and assesses how the capillarity is influenced by the absorption direction (vertical/horizontal), by the presence of interfaces and by pressure.

The adaptation and use of gamma ray equipment, by LFC-FEUP, has two main objectives described in detail below:

- Implement and develop a robust toolkit of gamma ray technique and analyses suited to the study of porous building materials and experimentally measure moisture content profiles, for use in the current project and for future projects;
- Obtain spatially resolved information. The gamma ray measurements are expected to provide detailed information regarding: (i) the differences in the absorption and drying processes of materials with distinct porosity and pore size distribution; (ii) the changes occurred in that processes due to the presence of soluble salts; (iii) the influence of different coatings, surface treatments and interfaces types.

The characterization of the different interface types together with the behaviour and influence of each in moisture transfer during absorption and drying processes between different common construction building materials is expected, facilitating the provision of detailed information regarding the following: the differences in the humidification and in the drying processes of materials with distinct porosity and pore size distribution; the influence of different coatings, surface treatments and interfaces types.

3.5.1 Measurement Principles

An atomic nucleus of a radioactive element can essentially emit three types of radiation: alpha particles, beta particles and gamma rays. Some basic differences in the nature of these elements must be taken into account. Alpha and beta particles have mass at rest although gamma radiation does not, allowing the last to travel in a vacuum at 300,000 km/s. Alpha particles show doubly ionized helium nuclei and, therefore, are positively charged; beta particles carry a negative charge, whereas gamma rays have no charge and therefore interact with the matter in a different way.

The operating principle of this measurement technique consists of the emission of gamma photons, in a small radioactive amount, which attenuate when passing through specific thick material, thus allowing the establishment of the following exponential relation between the number of emitted and received photons:

$$I = I_0 \cdot e^{-\mu\rho}x \qquad (3.1)$$

where I_0 is the emitted radiation intensity (counts), I is the received radiation intensity (counts), μ is the material attenuation coefficient (m^2/kg), ρ is the volumetric mass (kg/m^3) and x is the sample thickness (m).

The particles composing the gamma radiation present energy prevailing in electron volt, which gets divided according to a spectrum with one or more peaks depending on the nature of the radioactive source. Since gamma-ray measurement implies the use of monoenergetic radiation, it is necessary to consider only the photons whose energy is included between two levels near a peak.

The emission of photons, randomly performed, follows Poisson's law. Towards energy lower than 1000 keV, the total absorption of gamma radiation by the penetrated matter depends fundamentally on two mechanisms:

- Photoelectric effect, in which photons give up all their energy to an electron of a deep layer;
- Compton effect, which is defined by a photon deviation after the collision with a free electron, thus losing its energy which decreases proportionally to the diffusion angle.

Porous materials consist of their own solid skeleton and can also contain mater in the liquid phase. In addition, other materials involved in the specimens under study may exist. Therefore, the attenuation law will be expressed as follows:

$$I = I_0 \cdot e^{-\mu_i \, \rho_i \, x_i} \cdot e^{-\mu_0 \, \rho_0 \, d(1-\varepsilon)} \cdot e^{-\mu_w \, \rho_w \, d.\varepsilon.S}$$

$$\text{Envelope} \quad \text{Porous structure} \quad \text{Liquid phase} \tag{3.2}$$

where μ_i is the coefficient of envelope materials attenuation (m^2/kg), μ_0 is the coefficient of porous matrix attenuation (m^2/kg), μ_w is the coefficient of water attenuation (m^2/kg), x_i is the envelope i thickness (m), d is the thickness of porous material being studied (m), ρ_i is the volumetric mass of envelope materials (kg/m^3), ρ_0 is the volumetric mass of porous material being studied (kg/m^3), ρ_w is the volumetric mass of water (kg/m^3), ε is the porosity (%) and S is the saturation (%). Measuring with completely dry material, a value is obtained for the intensity of the radiation received, which is named I_0^*, so the expression presents the following configuration:

$$I = I_0 \cdot \exp(-\mu_w \cdot \rho_w \cdot d \cdot \varepsilon \cdot s) \tag{3.3}$$

The volumetric [$\theta = \varepsilon \cdot S$ (m^3/m^3)] or ponderal (W) moisture content can then be easily calculated, provided that the water attenuation coefficient is known, μ_w, as the volumetric mass of the dry material,

$$\theta = -\ln\left(\frac{I}{I_0^*}\right) \cdot \frac{I}{\mu_w \cdot \rho_w \cdot d} \quad \text{or} \quad w = \frac{\theta}{\rho_0} \tag{3.4}$$

Nielson [8], Kumaran et al. [10], Schwartz et al. [12], Freitas [13] and Qiu et al. [14] conducted the experimental determination of the water absorption coefficients— μ_w corresponding to an Americium 241's source, having achieved values close to 0.0191 (m^2/kg). If I_{sat} matches the intensity of the received radiation when it goes through a fully saturated sample, it may be considered that:

$$I_{sat} = I_0^* \cdot \exp(-\mu_w \cdot \rho_w \cdot d \cdot \varepsilon) \tag{3.5}$$

Solving the expression, the following relation, which allows calculating the saturation degree, is obtained:

$$S = \frac{\ln \frac{I}{I_o^*}}{\ln \frac{I_{sat}}{I_o^*}} \qquad (3.6)$$

Once the degree of saturation is known as well as the volumetric ($\theta = \varepsilon S$, m^3/m^3) and ponderal ($W = W_{sat} \cdot S$, in kg/kg) moisture contents, the results are easily obtained. When using Eq. (3.6) there will be less parameters prone to errors when comparing with the use of Eq. (3.4). However, there is great difficulty in obtaining saturation moisture content for some building materials as, for instance, for the cellular concrete. Therefore, Eq. (3.4) is used for concrete and Eq. (3.6) for the red brick material.

3.5.2   Setup

In this subsection, the parts of the HUMIGAMA.VF machine were presented in detail and some care for use.

3.5.2.1   Radioactive Source Collimator

The chosen radioactive source is Americium 241, which has already been used in other laboratories as reported above, with an activity of 100 mCi (3.7×109 Bq) and energy of 60 keV, showing thus the advantage of having a high contrast and a long half-life (458 years) (see Fig. 3.3).

Fig. 3.3 Radioactive source

Fig. 3.4 Detector inserted in protection

3.5.2.2 Detector

The detector, whose shielding and the measuring electronics were designed and assembled in Portugal by Laboratório Nacional de Engenharia e Tecnologia Industrial, was requested by the section of Civil Constructions—FEUP.

The detector is a crystal of sodium iodide (see Fig. 3.4), activated with thallium, and coupled to a photomultiplier with its preamplifier. The detector protection is intended to reduce the natural background of gamma radiation.

3.5.2.3 Measuring Electronics

The electronic unit contains the Power modules AT (see Fig. 3.5); Voltmeter (3 digits); Feeding system BT; Discriminator Disc 02; Counting rate meter; Timer; Counter (5 decades) and a panel with a display to show the content of the counter.

Fig. 3.5 Electronic unit

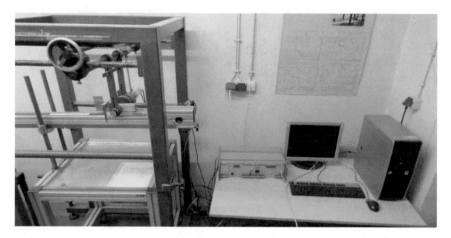

Fig. 3.6 General view of metal structure

3.5.2.4   Metal Support Structure

The metal support structure allows strict displacement of the detector and the radioactive source by presenting a collimator with a diameter of 5 mm, which minimizes the measurement errors inherent in the dispersion of the emitted particles (see Figs. 3.6 and 3.7).

3.5.2.5   Stabilizer

All electrical and electronic systems are designed and produced to function at maximum efficiency at a certain power supply, referred to as the nominal operating voltage. For different reasons, the voltage of the power distribution network does not remain constant, expressing significant fluctuations in relation to the nominal rate. This implies not only a loss of efficiency for appliances, even in terms of occasional inability of operation, but also a significant increase in the failure rate. The stabilizer is the electronic equipment responsible for correcting the voltage mains in order to provide the equipment with a stable and safe feeding. It also protects the equipment against most power grid problems. HUMIGAMA.VF has a stabilizer to which the equipment and the computer are connected.

3.5.2.6   Calculation of the Attenuation Coefficient

The calculation of the attenuation coefficient—μw was conducted by measuring the intensity of the emitted and received radiation after the crossing of parallelepiped plastic cells filled with water.

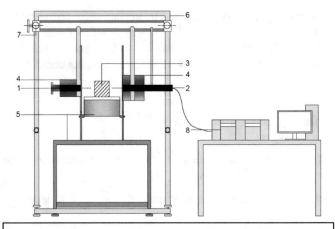

Legend:
1) Protected radioactive source
2) Protected detector
3) Sample
4) Collimators
5) Adjustable table to support the sample
6) Metal support
7) Crank that allows horizontal displacements
8) Measuring electronics (acquisition of results through software installed on the attached computer)

Fig. 3.7 Operating scheme of the device

"Dead time" of the detector—τ must be considered to get the actual counts, in particular for calculating the attenuation coefficients based on the following expression.

$$N = \frac{C}{t - C.\tau} - B \qquad (3.7)$$

where N is the Real counting's (counting's/s), t is the Counting time (s), τ is the Dead time (s/counting's), B is the Background radiation (counting's/s) and C is the Counting numbers (counting's/s).

3.5.2.7 Preparation of Test Specimens

The square cross-section of the test specimens in HUMIGAMA.VF should be not more than 100 mm × 100 mm. It is expected that the specimens are no more than 500 mm in terms of length.

The section of the specimens was defined taking into account that the phenomenon of on board connected to the waterproofing is more significant with section decrease. However, the use of the gamma ray attenuation method is not compatible with overly thick specimens, particularly in the case of samples of higher density.

3.5.3 Difficulty and Limitations

The recording of some practical difficulties related to this equipment's use is important for future users, namely:

- The maximum dimensions of the specimens that are 0.50 m length and 0.10 m width.
- The minimum distance between successive reading points should be 5 mm, due to the collimator features.
- The measurement errors inherent to the determination of the moisture content by attenuation of gamma radiation may come from the radioactive source whose photons emission is statistically random, from measurement electronics due to poor radiation monochromaticity, from the water attenuation coefficient, from the settlement of the absorbent material thickness, and from the development of the moisture content profile.
- Analyzing exclusively the random character of photons emission, the average error counting is $\pm\sqrt{n}$ so errors less than 0.005 require the number of counting's exceed 40,000, which corresponds to the change of times according to the materials density (10 s for a density of 500 kg/m^3 and 100 s for a density of 2000 kg/m^3).

3.5.4 Example of Moisture Content Profiles

Applying the experimental equipment described above, Freitas [13] calculated the imbibing water profiles of the cellular concrete and the red clay (see Figs. 3.8, 3.9 and 3.10). It is clear that the accuracy of the results is higher for the concrete, due to the fact that the density of both materials is different. The humidification of red clay is substantially faster than concrete. For example, in the red clay the wet front takes 3.5 h to reach the first 5 cm and 17 h in concrete.

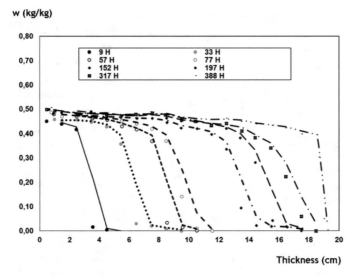

Fig. 3.8 Profiles of moisture content of cellular concrete in absorption

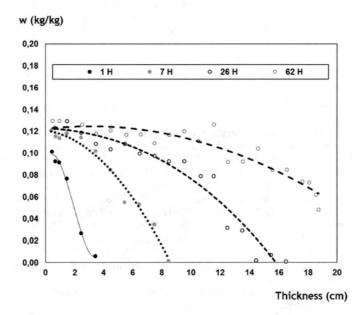

Fig. 3.9 Profiles of the moisture content of the red brick in absorption

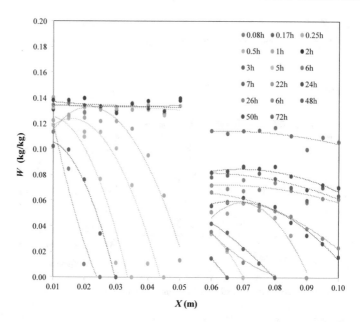

Fig. 3.10 Moisture content along the thickness of red brick with interface

References

1. L. Pel, Moisture transport in porous building materials. Ph.D. thesis, Technical University Eindhoven, The Netherlands (1995)
2. ISO 15148, Hygrothermal Performance of Building Materials and Products—Determination of Water Absorption Coefficient by Partial Immersion (2002)
3. R.J. Gummerson, C. Hall, W.D. Hoff, R. Hawkes, G.N. Holland, W.S. Moore, Unsaturated water flow within porous materials observed by NMR imaging. Nature **281**, 56–57 (1979)
4. R.F. Paetzold, G.A. Matzkanin, A. de los Santos, Surface soil water content measurement using pulsed nuclear magnetic resonance techniques—1. Soil Sci. Soc. Am. J. **49**(3), 537–540 (1985)
5. B. Guillot, R.D. Mountain, G. Birnbaum, Triplet dipoles in the absorption spectra of dense rare gas mixtures. I-Short range interactions. J. Chem. Phys. **90**(2), 650–662 (1989)
6. V. Mclane, C.L. Dunford, P.F. Rose, *Neutron Cross Sections: Neutron Cross Section Curves*, vol. 2 (Brookhaven National Laboratory, Upton, NY, USA, 1988)
7. H. Janssen, H. Derluyn, J. Carmeliet, Moisture transfer through mortar joints: a sharp-front analysis. Cem. Concr. Res. **42**, 1105–1112 (2012)
8. A. Nielsen, Gamma-ray—attenuation used for measuring the moisture content and homogeneity of porous concrete. Build. SCI **7**, 257–263 (1972)
9. M.K. Kumaran, M.A. Bomberg, A gamma-spectrometer for determination of density distribution and moisture distribution in building materials. National Research Council Canada, Division of Building Research (1985)
10. M.K. Kumaran et al., Moisture transport coefficient of pine from gamma ray absorption measurements, in *ASME Heat Transfer Division* (1989), pp. 179–183
11. F. Descamps, Continuum and discrete modelling of isothermal water and air transfer in porous media. Ph.D. thesis, Catholic University of Leuven, Leuven, Belgium (1997)

12. N.V. Schwartz, M. Bomberg, M.K. Kumaran, Water vapor transmission and moisture accumulation in polyurethane and polyisocyanurate foams, in *Water Vapor Transmission Through Building Materials and Systems: Mechanisms and Measurement* (ASTM International, USA, 1989)
13. V.P. de Freitas, Moisture transfer in building walls—interface phenomenon analysis. Ph.D. Thesis, Faculdade de Engenharia da Universidade do Porto, Porto, Portugal (1992)
14. X. Qiu, Moisture transport across interfaces between building materials. Ph.D. thesis, Concordia University, Montreal, Canada (2003)

Chapter 4
Interface Influence During the Wetting Process

4.1 Introduction

As it was previously said moisture damage is one of the most important pathological causes of building materials and components. A moisture measuring device based on a non-destructive method of gamma rays attenuation allows measuring this damage to broaden concepts in building physics related to moisture transfer; study the influence of the interface between layers in moisture transfer; analyse the influence of gravity on absorption of different types of building materials; study the kinetics of absorption of walls of one or more layers; analyse the importance of the temperature gradient in the movement of moisture; and calculate the coefficient of water diffusivity of some building materials, etc.

This chapter presents the results and produced in the analysis of an experimental campaign of water absorption in samples of red brick with different densities. In these experiments, the samples with and without joints at different height positions and different contact interface configurations were considered. Two experimental methods for the determination of the water absorbed during the wetting process were employed: the gravimetric method and gamma ray attenuation.

The results show a retardation of the wetting process due to the interfaces which is called hygric resistance. The samples with hydraulic contact interface with cement mortar present lower absorption rates than the samples with lime mortar. The influence of an air space between layers was also demonstrated. The air space interfaces significantly increase the capillary coefficients, as the distances from the contact with water also increase.

Intervening in old buildings requires extensive and objective knowledge. The multifaceted aspects of the work carried out on these buildings tend to include a growing number of disciplines, with a marked emphasis on those allowing for a greater understanding of the causes involved with the problems that affect them, as well as how to appropriately handle these causes.

© The Author(s), under exclusive license to Springer Nature Switzerland AG 2020   31
J. M. P. Q. Delgado et al., *Interface Influence on Moisture Transport in Building Components*, SpringerBriefs in Applied Sciences and Technology, https://doi.org/10.1007/978-3-030-30803-2_4

The study of moisture migration in the interior of materials and building construction components is of great importance for its behaviour characterization, especially regarding its durability, waterproofing, appearance degradation, and thermal problems.

For example, the study of the rising damp allowed for the investigation of a technique to solve this problem, which can already be used with the required revisions in order to treat building walls after a flood [1–5]. In Portugal, historical buildings stand near water courses with degraded walls due to the permanence of water. Moisture in buildings can have different origins of which rising damp is probably the most frequent. Floods are extreme occurrences but can permeate walls with large amounts of water. In conclusion, rising damp, because of its frequent occurrence, and floods, because of their consequential severity, both represent a high risk in terms of old building humidity problems.

The analysis of moisture migration in building materials and elements is crucial for knowledge of its behaviour. This behaviour involves durability, waterproofing, degradation and thermal performance of building wall. Generally, the wall consists of multiple layers, and thus the investigation of the moisture transfer implies knowledge of the continuity between layers. Freitas [6, 7] considered three different interfaces configurations:

- Perfect contact—when there is contact without interpenetration of both layers of porous structure;
- Hydraulic continuity—when there is interpenetration of both layers porous structure;
- Air space interface—when there is an air space of a few millimetres wide between layers.

In literature, although several studies concerning the liquid transport in multilayered porous structures exist, only a limited number of experimental values for the interface resistance in multilayered composites are found [8–11]. Qiu [9] shows that if the interface resistance is determined after capillary saturation of the first layer, the change in material properties could be neglected. The author analysed experimentally the liquid transport across the interface between aerated concrete and mortar. In addition, the author compared the experimental results with numerically obtained results using an interface resistance. Derluyn et al. [11, 12] considered an interface resistance as well as a change in mortar properties. The authors show that for dry cured mortar, a higher interface resistance was obtained, compared to the wet cured composite. Similar interface resistances analytically obtained by Janssen et al. [8] prove the validity of Derluyn's study.

In this chapter, new experimental values of water absorption in samples of red brick, with different densities, with and without interfaces, at varying positions, are presented, and an analysis of the three different interfaces configurations are reported and examined in detail.

4.2 Preparation of the Specimens

In the current subsection, the steps performed for specimen preparation (see Fig. 4.1) will be described. The red brick samples present different densities, type "A" with 1800 kg/m^3 and type "B" with 1600 kg/m^3. The samples area is 40 × 40 mm^2 (sectional area), with a height of 100 mm, for red brick type "A" and 50 × 50 mm^2 (sectional area), with a height of 100 mm, for red brick type "B".

The description of the typology, execution details, specimen preparation, among others factors is necessary to the experimental research, the aim of which is to discuss specimen production and the comparison of results, as well as the step-by-step preparation of the imbibition. These results are presented in the following chapters.

In this study, different interface configurations are analysed (see Table 4.1). The impact of an interface on moisture transport is evaluated by comparing the moisture flux of these samples with monolithic samples of the same material.

The tests include four different typologies: (i) monolithic block; (ii) perfect contact interface at 2, 5 and 7 cm; (iii) a block with cement and lime mortar interface at 2, 5 and 7 cm; (iv) an air space with 0.5 and 0.2 cm air space at 2, 5 and 7 cm; Fig. 4.1 illustrates the four types of tested blocks.

The first step consisted of cutting the specimens, as the sections were required to be quadratic. The cuts were performed in two different ceramics, A and B, giving origin to distinct specimens. At the end of the preparation, the specimens made of ceramics type "A" presented dimensions of 4 × 4 cm^2, while the ceramics type "B" presented a base of 5 × 5 cm^2, with variable height (Fig. 4.2). After being cut, the specimens were placed in a chamber for complete drying of the material.

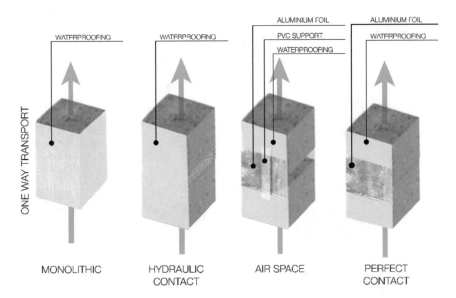

Fig. 4.1 Specimens used in the experimental campaign

Table 4.1 Interface configurations analysed

Material	Interface configurations analysed			
	Monolithic	Perfect contact	Hydraulic contact	Air space
Red brick type "A"		At 2 cm At 5 cm At 7 cm	Cement mortar at 2 cm Cement mortar at 5 cm Cement mortar at 7 cm Lime mortar at 2 cm Lime mortar at 5 cm Lime mortar at 7 cm	Cavity of 0.2 cm at 2 cm Cavity of 0.5 cm at 2 cm Cavity of 0.2 cm at 5 cm Cavity of 0.5 cm at 5 cm Cavity of 0.2 cm at 7 cm Cavity of 0.5 m at 7 cm
	1 sample × 5 unit	3 samples × 5 unit	6 samples × 5 unit	6 samples × 5 unit
Red brick type "B"		At 2 cm At 5 cm At 7 cm	Cement mortar at 2 cm Cement mortar at 5 cm Cement mortar at 7 cm Lime mortar at 2 cm Lime mortar at 5 cm Lime mortar at 7 cm	Cavity of 0.2 cm at 2 cm Cavity of 0.5 cm at 2 cm Cavity of 0.2 cm at 5 cm Cavity of 0.5 cm at 5 cm Cavity of 0.2 cm at 7 cm Cavity of 0.5 m at 7 cm
	1 sample × 5 unit	3 samples × 5 unit	6 samples × 5 unit	6 samples × 5 unit

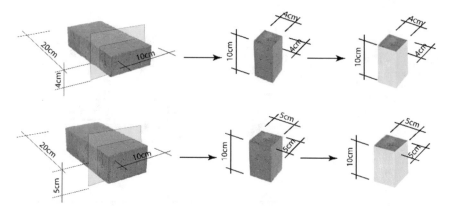

Fig. 4.2 Cutting process to imbibition

Fig. 4.3 Sealing process   IMBIBITION PROCESS

LATERAL SIDES
SEALED WITH
EPOXY RESIN

Following the in-chamber drying process, waterproofing was done on the lateral faces for the imbibition test as shown in Fig. 4.3, since the water/humidity absorption and evaporation process would be analysed exclusively for the unidirectional transfer.

For the waterproofing of the blocks, an epoxy resin, made up of two components, was used, mixing for 3 min. Each face was covered twice with the epoxy, first waterproofing the two opposite longitudinal faces and the two remaining faces after drying, which facilitates the process of specimen waterproofing. After drying the waterproofing material under normal environment temperature, the specimens were placed again in a chamber in a manner that allowed the mass of the materials to remain constant.

Specimen waterproofing is a process which requires caution in execution, since the faces which will be exposed to the water and air (top face) cannot come into contact with the waterproofing material. When these surfaces come into contact with the waterproofing material, sanding is necessary to eliminate all traces of the epoxy, which would impede the transfer of humidity within the block.

In order to waterproof the faces in perfect contact with the specimens, the faces were joined together with varying distances from the base of the specimen at 2, 5 and 7 cm (see Fig. 4.4), in order to compose a solid block. To guarantee a perfect union, i.e. avoiding air spaces between the blocks, wet-and-dry sandpaper was used. To assure the connection between the parts of the specimen and simultaneously avoid water evaporation through the interface, aluminium adhesive tape was used to anchor the blocks. When tested, despite being very similar to the monolithic specimens, a dissimilar behaviour in the perfect contact specimens was expected since there is a discontinuity in the porous structure.

In the specimens with an air space interface, as the name suggests, there is an air space between the union of the two blocks, with varying distances from the base of the specimens (at 2, 5 and 7 cm) (see Figs. 4.1 and 4.4) and with two different thickness (0.2 and 0.5 cm). In order to guarantee the thickness of this space between the blocks, movable PVC spacers of the same size were used, while fixed PVC pieces were used

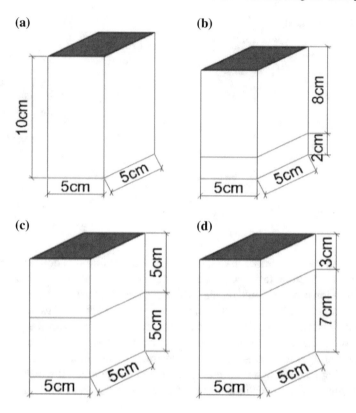

Fig. 4.4 Samples tested: **a** monolithic samples, **b** samples with interface at 2 cm, **c** samples with interface at 5 cm and **d** samples with interface at 7 cm

to join the two blocks by their lateral faces using a silicone bonding. The goal of the adapted material for the spacers was to guarantee the non-transfer of moisture in the place of fixation. After the silicone dried, the movable spacers were removed. An aluminium adhesive tape was then applied in order to guarantee the restriction of the airspace and impede the lateral evaporation, as can be seen in Fig. 4.1.

In the specimens with a mortar interface (cement and lime), with 2, 5 and 7 cm (Fig. 4.4) of distances from the sample base, a 1 cm thick layer was placed in the fresh state between the two blocks. The intention was to show that introducing materials with different properties between the blocks will produce a difference in moisture transfer at the interface. Since the mortars have a water absorption coefficient greater than block B and smaller than block A, it can be compared the influence.

After all the samples were taken, the imbibition test was carried out with the following procedure: (1) organization of the bench where the reservoir would be located; (2) reservoir levelling; (3) balance calibration; (4) marking the 5 mm level where the water will reach the samples; (5) filling the tank to the recommended level; (6) weighing the test pieces before the start of the tests; (7) placing the sample in the expected time; (8) weighing the specimen; (9) removal of the specimen from the receptacle; (10) elimination of excess water from the immersed surface with a damp cloth; (11) sample weighing; (12) weight annotation; (13) check the water level in the container.

Having carried out all the tests, the faces that were immersed in the water were measured, because not all the samples had the same measurement due to the undesirable imprecisions obtained in the cuts, which is able to influence the absorption relation/contact area.

4.3 Results and Discussion

4.3.1 Hygric Resistance

Moisture transport in multilayer porous materials, with discontinuities caused by the existence of an interface between the materials [7], was analysed by two experimental methods (gravimetric method and gamma ray attenuation). The existence of different interface types (perfect contact, hydraulic contact and air space interface) influence moisture transport, when compared to a monolithic porous element.

Since the goal is to address the effect of the interface in water absorption, the water resistance is measured immediately after the first changing point, which is the time interval of interest. The final measurement is performed by calculating each *Hygric Resistance* (*HR*). For the gravimetric method the calculation of the RH's was obtained through Eq. (4.1).

$$HR = \frac{\Delta M_w}{\Delta t} \tag{4.1}$$

where Δt (s) and ΔM_w (kg/m^2) are the variation of the time and water absorption immediately after the knee point, respectively.

As a consequence of these discontinuities in the porous structure of building materials, the interface causes a hygric resistance that limits the transport of moisture. The hygric resistance (*HR*) is represented by Fig. 4.5. This measurement is calculated experimentally (during a water absorption test) by the slope of the mass variation curve as function of time, after the knee point.

4.3.1.1 Gravimetric Method

Figures 4.6, 4.7, 4.8, 4.9, 4.10 and 4.11 display some examples of water absorption, showing differences between monolithic, perfect contact, hydraulic contact and air space interface curves, for the two materials. This experiment follows the standard ISO 15148 [13].

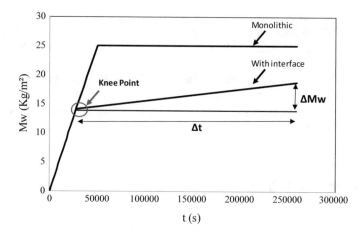

Fig. 4.5 Hygric Resistance (*HR*) in gravimetric method

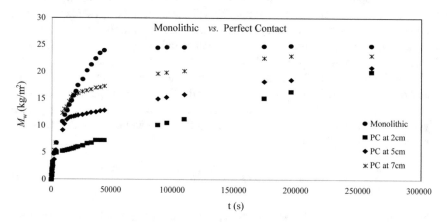

Fig. 4.6 Comparison between a monolithic sample and a sample with perfect contact, for red brick type "A"

Figures 4.8 and 4.9 show that the water absorption for the samples with perfect contact, before the interface, exhibit similar behaviour to the monolithic samples. When the water front reaches the interface (at 2 cm 5 cm or 7 cm in height), a water resistance is verified, reducing the absorption rate.

Figures 4.10 and 4.11 indicate that the water absorption for the samples with hydraulic contact, before the interface, exhibits behaviour similar to the monolithic samples. When water reaches the interface (at 2 cm 5 cm or 7 cm in height), a water resistance is verified, reducing the absorption rate.

As in the previous examples, Figs. 4.10 and 4.11 demonstrate that water absorption for the samples with an air space before the interface exhibits a behaviour similar to the monolithic samples. When water reaches the interface (at 2 cm 5 cm or 7 cm in

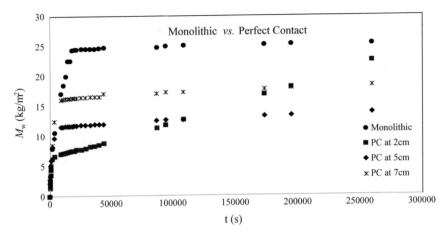

Fig. 4.7 Comparison between a monolithic sample and a sample with perfect contact, for red brick type "B"

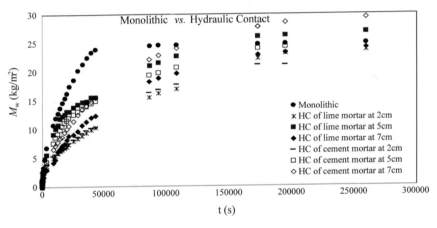

Fig. 4.8 Comparison between a monolithic sample and a sample with hydraulic contact, for red brick type "A"

height), a water resistance is verified, and after that it remains almost an horizontal which means that the absorption processes almost stopped.

Table 4.2 summarizes the experimental values of hygric resistance obtained for the perfect contact, hydraulic contact and air space interface tested samples.

The following figures presents a comparison between the obtained values from this study with those obtained from the literature. Figure 4.12 compares the obtained value of the hygric resistance for perfect contact, which is similar to that found in [6, 14].

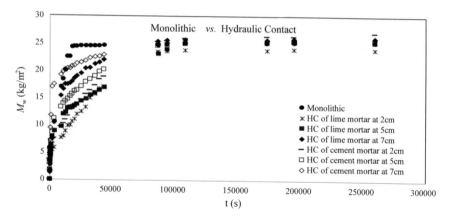

Fig. 4.9 Comparison between a monolithic sample and a sample with hydraulic contact, for red brick type "B"

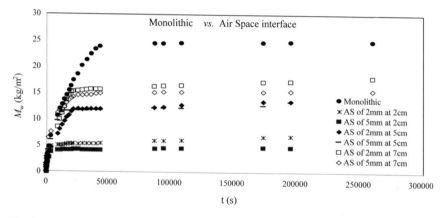

Fig. 4.10 Comparison between a monolithic sample and a sample with air space interface, for red brick type "A"

Figure 4.13 show that the obtained values of hygric resistance for air space here determined are similar to that found by Freitas [6] and Guimarães et al. [14]. Figure 4.14 show that the obtained curvature shape for the specimens with mortar cement, is similar to that found in [12, 14, 15].

The samples with mortar interface, as previously reported, are produced by placing this mortar in the fresh state. As this mortar is in the fresh state the bottom block absorbs water and cement from the mortar (Fig. 4.15). Figure 4.15 presents the new approach presented below in detail. In this analysis the capillary coefficient of monolithic bricks was compared with the capillary coefficient of bricks with hydraulic interface (Fig. 4.15a), before to reach the interface (Fig. 4.15c).

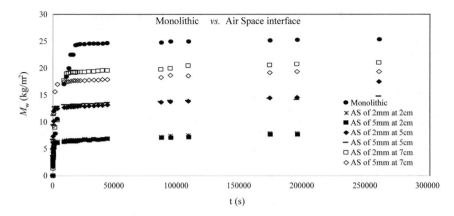

Fig. 4.11 Comparison between a monolithic sample and a sample with air space interface, for red brick type "B"

The coefficients of capillarity obtained in samples with hydraulic contact interface, the results obtained with monolithic bricks (A and B) and by other authors, presented in literature (Fig. 4.16), are described in Table 4.3.

Depraetere et al. [15] and Rego [16] observed that there is a change in the capillarity coefficient before the corner to the cement mortar interface. This occurs due to the fact that, when one employs a fresh mortar between the dry ceramic blocks, there is an absorption of water and cement which is in the mortar, making the capillarity coefficient of the brick decrease when compared with the isolated brick. Derluyn et al. [12], for example, observed a different behaviour, where in the same conditions there was an increase of the capillarity coefficient of the tested blocks. Apparently, it is possible to conclude from these different results from literature is that there is a change in capillarity coefficient in the blocks which are in contact with the mortar.

In conclusion, from this research, it is possible to ascertain the following conclusions:

- The water absorption rate is related to the position of interface height, increasing by more than 10% for red brick type "B" and above 8% for red brick type "A". This phenomenon was not observed in samples with perfect contact interface;
- The difference between the monolithic samples tested and the samples with hydraulic contact (cement mortar and lime mortar) is the decrease of the absorption rate observed in the samples with hydraulic contact;
- The samples with hydraulic contact interface with cement mortar present lower absorption rates than the samples with lime mortar;
- The situation where absorption reaches the value of the monolithic test result is with an interface of hydraulic contact with lime mortar;
- The highest value of hygric resistance was obtained with an interface of hydraulic contact with cement mortar;
- The two materials tested with air space interface show an initial constant absorption rate and, when the humidity reaches the interface, a very slow absorption;

Table 4.2 Values of hygric resistance for the perfect contact, hydraulic contact and air space interface samples

Material	Sample/(interface type)	Hygric resistance (kg/m²s)	Standard deviation	Coefficient variation (%)
Red brick type "A"	Perfect contact (2 cm)	6.6×10^{-5}	7.7×10^{-6}	11.5
	Perfect contact (5 cm)	4.3×10^{-5}	12.0×10^{-6}	27.2
	Perfect contact (7 cm)	2.4×10^{-5}	3.8×10^{-6}	15.6
	Hydraulic contact (cement mortar at 2 cm)	9.7×10^{-5}	5.2×10^{-6}	5.3
	Hydraulic contact (cement mortar at 5 cm)	7.1×10^{-5}	19.0×10^{-6}	26.2
	Hydraulic contact (cement mortar at 7 cm)	4.2×10^{-5}	0.6×10^{-6}	1.6
	Hydraulic contact (lime mortar at 2 cm)	7.9×10^{-5}	5.5×10^{-6}	7.0
	Hydraulic contact (lime mortar at 5 cm)	5.8×10^{-5}	2.2×10^{-6}	3.7
	Hydraulic contact (lime mortar at 7 cm)	5.4×10^{-5}	5.8×10^{-6}	10.8
	Air space 2 mm at 2 cm	0.8×10^{-5}	1.2×10^{-6}	14.0
	Air space 5 mm at 2 cm	0.4×10^{-5}	1.2×10^{-6}	28.2
	Air space 2 mm at 5 cm	0.9×10^{-5}	0.3×10^{-6}	3.6
	Air space 5 mm at 5 cm	0.8×10^{-5}	0.2×10^{-6}	2.3
	Air space 2 mm at 7 cm	0.9×10^{-5}	1.4×10^{-6}	15.1
	Air space 5 mm at 7 cm	0.4×10^{-5}	0.6×10^{-6}	16.1

(continued)

Table 4.2 (continued)

Material	Sample/(interface type)	Hygric resistance (kg/m^2 s)	Standard deviation	Coefficient variation (%)
Red brick type "B"	Perfect contact (2 cm)	7.3×10^{-5}	14.0×10^{-6}	19.4
	Perfect contact (5 cm)	5.0×10^{-5}	3.9×10^{-6}	7.9
	Perfect contact (7 cm)	0.9×10^{-5}	0.3×10^{-6}	2.7
	Hydraulic contact (cement mortar at 2 cm)	7.9×10^{-5}	1.7×10^{-6}	2.1
	Hydraulic contact (cement mortar at 5 cm)	4.8×10^{-5}	4.7×10^{-6}	9.7
	Hydraulic contact (cement mortar at 7 cm)	3.2×10^{-5}	3.5×10^{-6}	11.1
	Hydraulic contact (lime mortar at 2 cm)	7.9×10^{-5}	11.0×10^{-6}	14.1
	Hydraulic contact (lime mortar at 5 cm)	4.2×10^{-5}	1.6×10^{-6}	3.9
	Hydraulic contact (lime mortar at 7 cm)	4.2×10^{-5}	4.2×10^{-6}	9.8
	Air space 2 mm at 2 cm	0.7×10^{-5}	0.1×10^{-6}	1.8
	Air space 5 mm at 2 cm	0.8×10^{-5}	0.8×10^{-6}	10.0
	Air space 2 mm at 5 cm	1.0×10^{-5}	0.1×10^{-6}	9.4
	Air space 5 mm at 5 cm	0.7×10^{-5}	0.8×10^{-6}	12.3
	Air space 2 mm at 7 cm	0.7×10^{-5}	0.9×10^{-6}	13.9
	Air space 5 mm at 7 cm	0.9×10^{-5}	1.6×10^{-6}	18.9

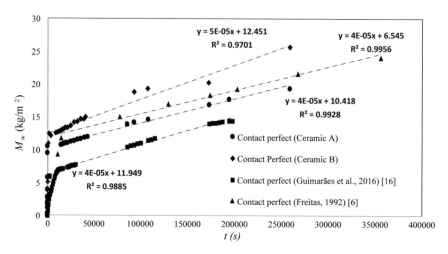

Fig. 4.12 Comparison between the obtained values in perfect contact with that obtained from the literature

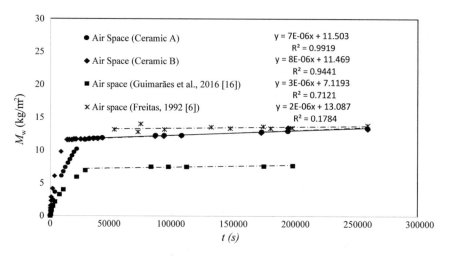

Fig. 4.13 Comparison between the obtained values in space air with that obtained from the literature

- The air space interfaces increase significantly capillary coefficients, as the distances from the contact with water increase;
- In perfect contact interfaces, some values were discarded due to the fact that after the knee, the absorption curve assumes constant values. However, a perfect physical contact is difficult to achieve. In fact, the interface of an imperfect contact is characterized by a combination of air contact and natural contact conditions.

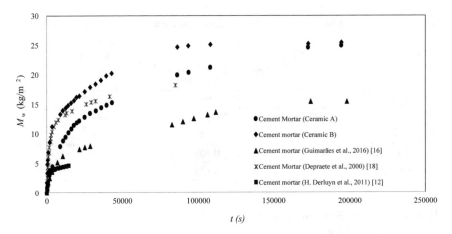

Fig. 4.14 Comparison between the obtained values with mortar cement with that obtained from the literature

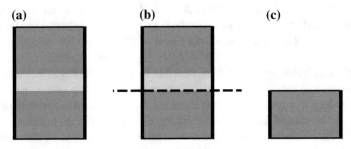

Fig. 4.15 **a** Sample with interface; **b** separação do tijolo e interface; **c** the blocks before and after the employment of a mortar cement interface

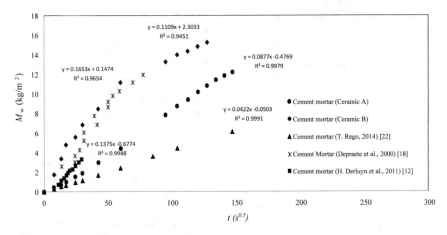

Fig. 4.16 Values of the capillarity coefficients for bricks before and after the employment of the cement mortar interface

Table 4.3 Comparison of the values for the capillarity coefficients for the monolithic bricks before of cement mortar interface

Material	A (kg m^{-2} s$^{-0.5}$) (Monolithic brick)	A (kg m^{-2} s$^{-0.5}$) (brick before the interface)	Influence interface (%)
Ceramic A	0.100	0.087	−13
Ceramic B	0.190	0.110	−42
Rego [16]	0.068	0.042	−38
Depraetere et al. [15]	0.184	0.165	−10
Derluyn et al. [12]	0.116	0.138	18

4.3.1.2 Gamma Ray Attenuation

A moisture measuring device based on the non-destructive method of gamma ray attenuation was used to study moisture transfer and to obtain moisture content profiles in two samples of red brick—types "A" and "B"—with different sectional area, density and interfaces types: contact perfect, air space and hydraulic contact.

Figures 4.17, 4.18, 4.19, 4.20, 4.21 and 4.22 present the moisture content profiles along the thickness of the samples tested with gamma ray attenuation (HUMIGAMA.VF equipment). It is possible to observe that the existence of an interface between two layers causes a water resistance with a consequent delay in imbibition. In a few minutes the first layer reaches a moisture content close to saturation but the second layer, because of the different types of interface, behaves differently. In the perfect contact, the second layer is conditioned by the water resistance as occurs with lime mortar. With the air space, the second layer converges to the hygroscopic humidity and with the cement mortar, the interface prevents the transport of moisture functioning as a hygric cut.

There are some results that should be repeated in a future work. For example, in Fig. 4.18, where the discontinuity is observed, a different concavity of some curves obtained was expected. In addition, in Figs. 4.17, 4.18, 4.19, 4.20, 4.21 and 4.22, it is possible to observe some outliers that were not expected. However, the heterogeneity of the materials tested can explain the existence of the outliers observed.

When comparing the profiles, at the same point of analysis, of the perfect contact interface and the air space interface it is possible to conclude that:

- For ceramics "A":

  - The blocks with a 2 cm interface, the water front, when in perfect contact, reached the 5 cm height at the end of 3 h, while the blocks with an air space interface, the same results only occurred after 72 h;
  - The blocks with a 5 cm interface, the water front, when in perfect contact, reached the entire specimen (10 cm) within 24 h, while the blocks with an air space interface, after 72 h the water front was still found at 8 cm;

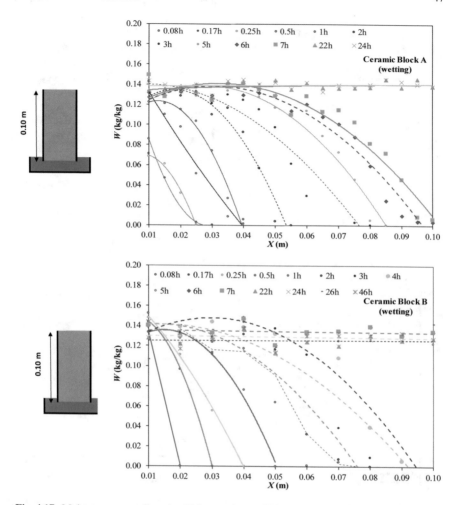

Fig. 4.17 Moisture content along the thickness of monolithic samples of red brick type "A and B"

- The blocks with a 7 cm interface, when in perfect contact, total saturation occurred within 26 h, and in the case of the air space interface, the water front reached the 10 cm section of the specimen after 72 h.

- For ceramics "B":

  - For blocks with 2 cm interface, the water front, when in perfect contact, reached the 8 cm section after 48 h. On the other hand, for the space air interface, after 48 h the water front was found at the 5.5 cm section;

  - For the blocks with 5 cm and 7 cm interface, the water front in both sides (perfect contact and space air interfaces) presented a similar behaviour up to 72 h of test duration.

(a) Contact perfect   (b) Air space

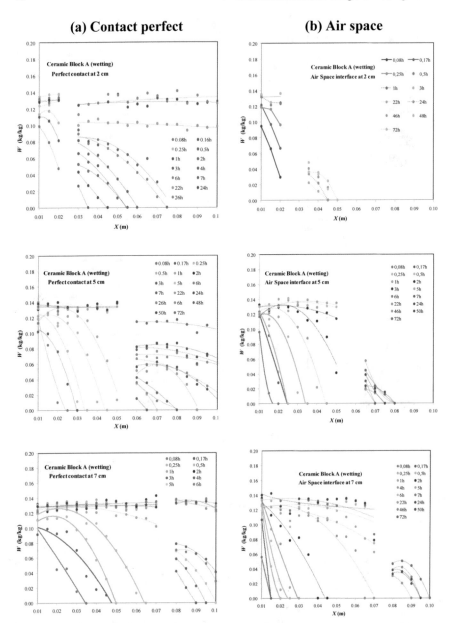

Fig. 4.18 Moisture content along the thickness of red brick type "A" samples with interface: **a** perfect contact and **b** air space at 2, 5 and 7 cm

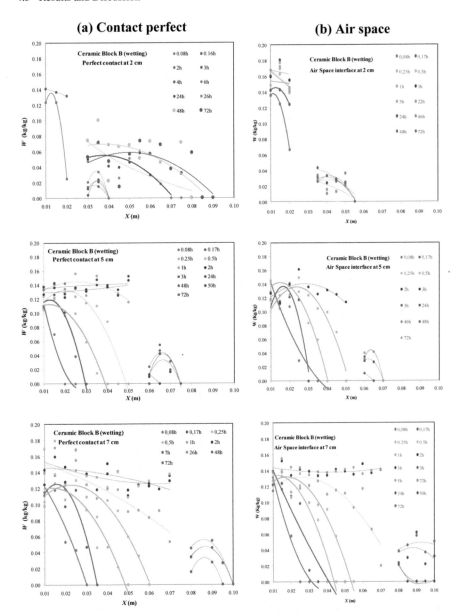

Fig. 4.19 Moisture content along the thickness of red brick type "B" samples with interface: **a** perfect contact and **b** air space at 2, 5 and 7 cm

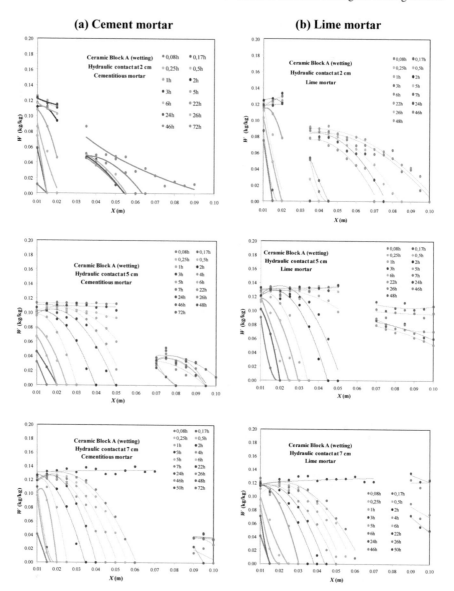

Fig. 4.20 Moisture content along the thickness of red brick type "A" samples with interface: **a** interface of cement mortar and **b** lime mortar at 2, 5 and 7 cm

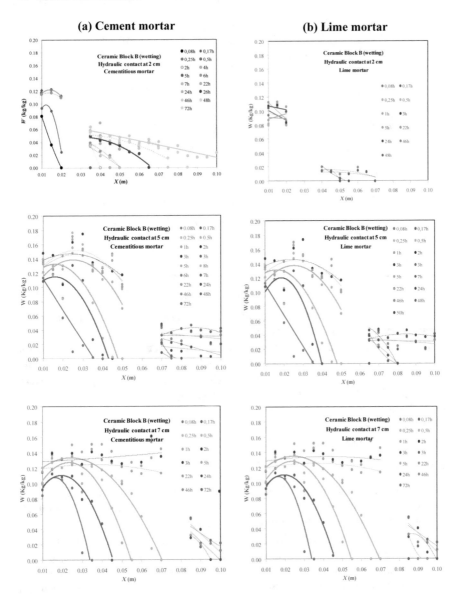

Fig. 4.21 Moisture content along the thickness of red brick type "B" samples with interface: **a** interface of cement mortar and **b** lime mortar at 2, 5 and 7 cm

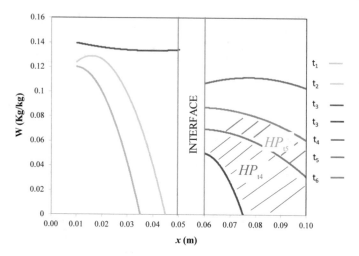

Fig. 4.22 Example of moisture profile curves during an imbibition process

Tables 4.4 and 4.5 present a comparison between the velocities of moisture front progression of different interfaces (perfect contact and air space).

From Tables 4.4 and 4.5, it is possible to conclude that the velocity of progression of the humidity surface, after the transition between interfaces. In, ceramic A, the moisture transfer with air space interface is clearly slower than with perfect contact. In ceramic B the two interfaces present a very similar velocity of progression of the humidity surface.

When comparing the two profile types in hydraulic contact (mortar cement and lime mortar), the following stands out:

- **For ceramics "A":**

  - In the case of the blocks with a 2 cm interface, the water front in the specimens with cement mortar, after 72 h, did not reach the 10 cm level. On the other hand, with lime mortar the 10 cm level was reached after 46 h;
  - In the case of the blocks with a 5 cm interface, the water front of the specimens with cement mortar reach the entire specimen after 48 h and until 72 h, while with the lime mortar, the same occurred before the 22-h time mark;
  - In the case of the blocks with a 7 cm interface, the water front of the specimens with mortar cement reach the 10 cm level within 26 h and with lime mortar before the 22-h time mark.

- **For ceramics "B":**

  - In the case of the blocks with a 2 cm interface, the water front in the specimens with mortar cement reached the 8 cm level between the 48 h and the 72-h time step. On the other hand, with the lime mortar, this level was reached after 48 h;

Table 4.4 Comparison of the velocity (m/s) of progression of the moisture front after the transition from the perfect contact interface to the air space interface (ceramics "A")

Interface type (ceramic A)	Perfect contact			Air space		
Interface level (cm)	2	5	7	2	5	7
Time interval after the interface transition (h)	3	24	26	72	72	72
Measured distance after the interface transition (cm)	5	5	3	5	3	3
Velocity (m/s) of progression after the interface transition	4.6×10^{-6}	5.7×10^{-7}	3.2×10^{-7}	1.9×10^{-7}	1.1×10^{-7}	1.1×10^{-7}

Table 4.5 Comparison of the velocity (m/s) of progression of the moisture front after the transition from the perfect contact interface and the air space interface (ceramic "B")

Interface type (ceramic B)	Perfect contact			Air space		
Interface level (cm)	2	5	7	2	5	7
Time interval after the interface transition (h)	48	72	72	48	72	72
Measured distance after the interface transition (cm)	6	5	3	3.5	5	3
Velocity (m/s) of progression after the interface transition	3.5×10^{-7}	1.9×10^{-7}	1.1×10^{-7}	2.0×10^{-7}	1.9×10^{-7}	1.1×10^{-7}

- In the case of the blocks with a 5 cm interface, the water front of the specimens with mortar cement reached the entire surface extension between 24 and 48 h, while in the specimens with lime mortar the same occurred after 22 h;
- In the case of the blocks with a 7 cm interface, the water front of the specimens with cement mortar reached the 10 cm level within 46 h. For the specimens with lime mortar, the same occurred before the 22-h time step.

Tables 4.6 and 4.7 present a comparison between the velocities of moisture surface progression of different hydraulic interfaces (cement lime and mortar lime).

From Tables 4.6 and 4.7, it is possible to conclude that the velocity of moisture surface progress, after the transition interface in the specimens with cement mortar is within 1.8×10^{-7} m/s and 3.5×10^{-7} m/s, while in the specimens with lime mortar is within 3.5×10^{-7} m/s and 6.3×10^{-7} m/s. In the HUMIGAMA.VF humidity profiles, the transfer of water from the cement mortar interface is slower than in the lime mortar interface.

Finally, it is presented the hygric resistance values (HR) determined by gamma ray method. Figure 4.22 shows an example of the moisture profiles obtained during the imbibition process, for different times. It was used the moisture profiles measured with HUMIGAMA.VF device, and the calculation of HP_t by the following equation

$$HR_t = \frac{\left(w_i + 2 \sum_{j=i+1}^{n-1} w_j\right) + w_n}{2} (x_{i+1} - x_i) \qquad (4.2)$$

As the mathematical functions that describe the moisture profiles curves are unknown, we used the experimental points associated to each curve to determine the integral by trapezoidal (Eq. 4.2) rule in order to avoid the total amount of water that passes through the interface (i.e., the amount of water that passes at a given time minus the amount of water passed at the previous instant is calculated).

Table 4.8 presents the RH values obtained for by Gamma-ray method.

Finally, it is interesting to present a comparison between the gravimetric method and gamma ray attenuation technique:

- In monolithic samples, water is continuously absorbed to the maximum, i.e. until the top surface is reached;
- It is possible to observe that the damp front velocity presents a relation with water absorption coefficient (A_w). Comparing the two ceramic brick samples ("A" and "B") with the different dimensions ($4 \times 4 \times 10$ cm^3 and $5 \times 5 \times 10$ cm^3) and considering the damp front obtained with ceramic brick type "B", which displays a higher value of water absorption coefficient, reaches the face in contact with the environment (at 10 cm) in a shorter time period. On the other hand, it should be stated that the effect of the wall´s thickness has to be carefully considered. In fact, the damp front increases with the wall thickness, following approximately a function of the square-root of this thickness [4, 5];
- In samples with perfect contact, the absorbed water is lower than in monolithic samples and the discontinuity is clearly identified. It is possible to observe that for

Table 4.6 Comparison of the velocity (m/s) of moisture surface progression of different hydraulic interfaces (cement mortar and lime mortar), for ceramic "A"

Interface type (ceramic B)	Cement mortar			Lime mortar		
Interface level (cm)	2	5	7	2	5	7
Time interval after the interface transition (h)	72	48–72	26	46	22	22
Measured distance after the interface transition (cm)	6	5	3	8	5	3
Velocity (m/s) of progression after the interface transition	2.3×10^{-7}	$2.9\text{–}1.9 \times 10^{-7}$	3.2×10^{-7}	4.8×10^{-7}	6.3×10^{-7}	3.8×10^{-7}

Table 4.7 Comparison of the speed (m/s) of progression of the humidity surface after the transition from the cement mortar to the lime mortar (ceramics "B")

Interface type (ceramic B)	Cement mortar			Lime mortar		
Interface level (cm)	2	5	7	2	5	7
Time interval after the interface transition (h)	48–72	24–48	46	48	22	22
Measured distance after the interface transition (cm)	6	5	3	6	5	3
Velocity (m/s) of progression after the interface transition	3.5–2.3×10^{-7}	5.6–2.9×10^{-7}	1.8×10^{-7}	3.5×10^{-7}	6.3×10^{-7}	3.8×10^{-7}

Table 4.8 Hygric resistance values determined gamma-ray method

Material	Sample/(interface type)	Gamma-ray method
Red brick type "A"	Perfect contact (2 cm)	8.7×10^{-5}
	Perfect contact (5 cm)	3.9×10^{-5}
	Perfect contact (7 cm)	1.6×10^{-5}
	Hydraulic contact (cement mortar at 2 cm)	1.8×10^{-5}
	Hydraulic contact (cement mortar at 5 cm)	4.4×10^{-5}
	Hydraulic contact (cement mortar at 7 cm)	7.6×10^{-5}
	Hydraulic contact (lime mortar at 2 cm)	4.5×10^{-5}
	Hydraulic contact (lime mortar at 5 cm)	2.2×10^{-5}
	Hydraulic contact (lime mortar at 7 cm)	1.6×10^{-5}
	Air space 2 mm at 2 cm	0.6×10^{-5}
	Air space 5 mm at 2 cm	0.5×10^{-5}
	Air space 2 mm at 5 cm	0.8×10^{-5}
	Air space 5 mm at 5 cm	1.4×10^{-5}
	Air space 2 mm at 7 cm	0.7×10^{-5}
	Air space 5 mm at 7 cm	0.2×10^{-5}
Red brick type "B"	Perfect contact (2 cm)	4.1×10^{-5}
	Perfect contact (5 cm)	0.7×10^{-5}
	Perfect contact (7 cm)	1.8×10^{-5}
	Hydraulic contact (cement mortar at 2 cm)	3.5×10^{-5}
	Hydraulic contact (cement mortar at 5 cm)	2.9×10^{-5}
	Hydraulic contact (cement mortar at 7 cm)	5.1×10^{-5}
	Hydraulic contact (lime mortar at 2 cm)	4.1×10^{-5}
	Hydraulic contact (lime mortar at 5 cm)	3.3×10^{-5}
	Hydraulic contact (lime mortar at 7 cm)	3.2×10^{-5}
	Air space 2 mm at 2 cm	1.1×10^{-5}
	Air space 5 mm at 2 cm	1.1×10^{-5}
	Air space 2 mm at 5 cm	1.3×10^{-5}
	Air space 5 mm at 5 cm	–
	Air space 2 mm at 7 cm	3.9×10^{-5}
	Air space 5 mm at 7 cm	0.7×10^{-5}

higher positions of the interface, the water absorbed is greater. This was quantified by the gravimetric method and the results of the gamma ray attenuation technique showed a higher water absorption in the first layer;

- The results with hydraulic contact interface (cement mortar) samples present lower absorption rate than the samples with lime mortar, for all techniques used. The interface of hydraulic contact of lime mortar is the only situation where the final absorption reaches the value of the monolithic test results;
- The results of the air space interface samples showed, after the interface, that the water absorbed is almost zero in gravimetric method. In the gamma ray attenuation technique, it stays close to the equilibrium water content, which means the same.

4.3.2 Synthesis

This chapter presents the results of experimental campaign and a critical analysis of water absorption processes. This analysis presented is comprised of samples of red brick with different densities, with and without joints at different height positions and different contact interface configurations, using two experimental methods: gravimetric method and gamma ray attenuation. A moisture measuring device based on the non-destructive method of gamma ray attenuation was used to study the moisture transfer. In this experiment, moisture content profiles in samples of red brick, type "A" and "B", was obtained with different sectional area and density.

The results show that when the moisture reaches the interface there is a slowing of the wetting process due to the interfaces hygric resistance. The samples with hydraulic contact interface cement mortar present lower absorption rates than the samples with lime mortar. The only situation where final absorption reaches the value of the monolithic test result was with an interface of hydraulic contact lime mortar and the highest value of hygric resistance was obtained with an interface of hydraulic contact of cement mortar.

It was possible to observe the influence of air space between layers. If the layers of consolidated materials are separated by an air space interface, an initial constant absorption rate was observed instead of a very slow absorption when the water reaches the interface. The air space interfaces increase the coefficients of capillary significantly, as the distances from the contact with water increase.

References

1. J.M.P.Q. Delgado, A.S. Guimarães, V.P. de Freitas, I. Antepara, V. Kočí, R. Černý, Salt damage and rising damp treatment in building structures. Adv. Mater. Sci. Eng. **2016**, Article number ID 1280894, 13 pages (2016)
2. E. Barreira, R. Almeida, J.M.P.Q. Delgado, Infrared thermography for assessing moisture related phenomena in building components. Constr. Build. Mater. **110**, 251–269 (2016)

3. A.S. Guimarães, J.M.P.Q. Delgado, V.P. Freitas, Rising damp in walls: evaluation of the level achieved by the damp front. J. Build. Phys. **37**, 6–27 (2013)
4. A.S. Guimarães, J.M.P.Q. Delgado, V.P. Freitas, Rising damp in building walls: the wall base ventilation system. Heat Mass Transf. **48**, 2079–2085 (2012)
5. V.P. Freitas, A.S. Guimarães, J.M.P.Q. Delgado, The HUMIVENT device for rising damp treatment. Recent Patents Eng. **5**, 233–240 (2011)
6. V.P. de Freitas, Moisture transfer in building walls—interface phenomenon analysis. Ph.D. thesis, Faculdade de Engenharia da Universidade do Porto, Porto, Portugal (1992)
7. V.P. de Freitas, V. Abrantes, P. Crausse, Moisture migration in building walls: analysis of the interface phenomena. Build. Environ. **31**, 99–108 (1996)
8. H. Janssen, H. Derluyn, J. Carmeliet, Moisture transfer through mortar joints: a sharp-front analysis. Cem. Concr. Res. **42**, 1105–1112 (2012)
9. X. Qiu, Moisture transport across interfaces between building materials. Ph.D. thesis, Concordia University, Montreal, Canada (2003)
10. P. Mukhopadhyaya, P. Goudreau, K. Kumaran, N. Normandin, Effect of surface temperature on water absorption coefficient of building materials. J. Therm. Envelope Build. Sci. **26**, 179–195 (2002)
11. S. Roels et al., Interlaboratory comparison of hygric properties of porous building materials. J. Therm. Environ. Build. **27**(4), 307–325 (2004)
12. H. Derluyn, H. Janssen, J. Carmeliet, Influence of the nature of interfaces on the capillary transport in layered materials. Constr. Build. Mater. **25**(9), 3685–3693 (2011)
13. ISO 15148, Hygrothermal Performance of Building Materials and Products—Determination of Water Absorption Coefficient by Partial Immersion (2002)
14. A.S. Guimarães, J.M.P.Q. Delgado, T. Rego, V.P. de Freitas, The effect of salt solutions in the capillarity absorption coefficient of red brick samples. Defect Diffus. Forum **369**, 168–172 (2016)
15. W. Depraetere, J. Carmeliet, H. Hens, Moisture transfer at interfaces of porous materials: measurements and simulations. PRO **12**, 249–259 (2000)
16. T. Rego, Efeito de Soluções Aquosas Salinas nos Processos de Embebição de Paredes com Múltiplas Camadas. M.Sc. thesis, Faculdade de Engenharia da Universidade do Porto, Portugal (2014)

Chapter 5
Conclusions

Moisture transport through a building envelope normally involves interface phenomena, i.e., moisture transport across interfaces between building materials. Therefore, the knowledge of the interface phenomena is essential for the prediction of moisture behaviour in a building envelope. Most hygrothermal models treat materials as individual layers in perfect hydraulic contact, i.e., the interface has no effect on the moisture transport. However, in practice, this might not always be true. Therefore, to appropriately evaluate the performance of a building envelope on moisture transport that lead to building envelope design guidelines, it is imperative to obtain a good understanding of the interface phenomena. The most common types of interfaces are: "Hydraulic contact" when there is an interpenetration of both layer's porous structure; "Perfect contact" when there is a contact without interpenetration and "Air space" between layers when there is an air box of a few millimetres wide between the layer's porous structure.

A brief description of the most important experimental transient methods—Gravimetric Method, Nuclear Magnetic Resonance (NMR) method, X-Ray Analysis and Gamma Ray attenuation method—to determine moisture content profiles was presented. The gravitational method and the attenuation of gamma ray's, HUMIGAMA.VF, were the methods chosen to obtain the experimental results presented in this research. The first one is an easy-to-use method worldwide and the second is a machine that was developed in the Laboratory of Building Physics (LFC), i.e., the laboratory where the research was conducted.

Finally, a detailed experimental campaign and a critical analysis of water absorption in samples of red brick with different densities, with and without joints at different height positions and different contact interface configurations, using two experimental methods (gravimetric method and gamma ray's attenuation), was presented. A moisture measuring device based on non-destructive method of gamma ray's attenuation was used to study the moisture transfer and obtained moisture content profiles in two samples of red brick, types "A" and "B", with different sectional area and density.

© The Author(s), under exclusive license to Springer Nature Switzerland AG 2020   61
J. M. P. Q. Delgado et al., *Interface Influence on Moisture Transport*
in Building Components, SpringerBriefs in Applied Sciences and Technology,
https://doi.org/10.1007/978-3-030-30803-2_5

The results show that when the moisture reaches the interface there is a slowing of the wetting process due to the interfaces hygric resistance. The samples with hydraulic contact interface (cement mortar) present lower absorption rate than the samples with lime mortar. The only situation where final absorption reaches the value of the monolithic test result is with an interface of hydraulic contact (lime mortar) and the highest value of hygric resistance was obtained with an interface of hydraulic contact (cement mortar).

It was possible to observe the influence of air space between layers. If the layers of consolidated materials are separated by an air space interface an initial constant absorption rate instead a very slow absorption when the water reaches the interface were observed. The air space interfaces increase the coefficients of capillary significantly, as the distances from the contact with water increase.

Printed in the United States
By Bookmasters